Gardening with Perennials

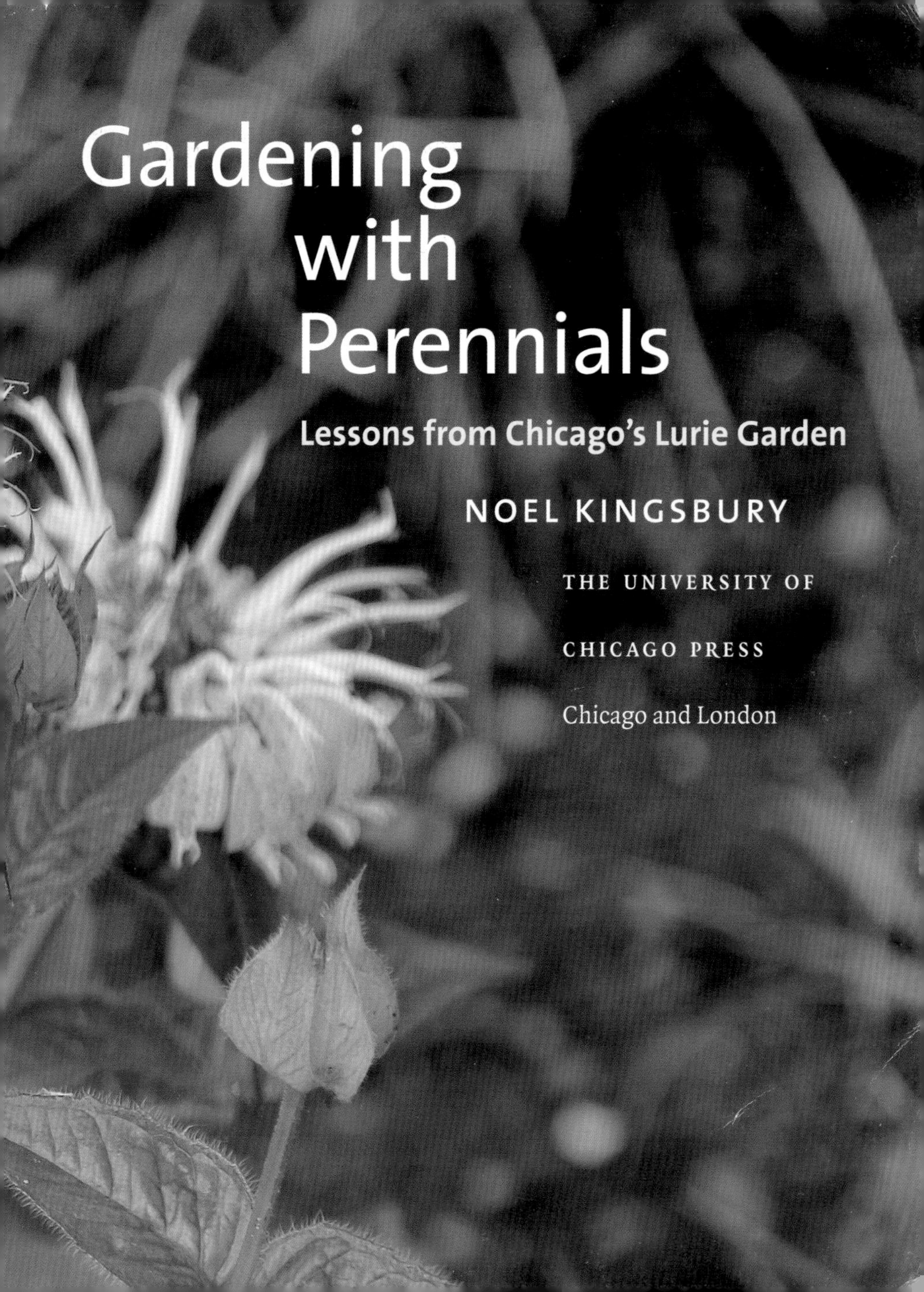

Gardening with Perennials

Lessons from Chicago's Lurie Garden

NOEL KINGSBURY

THE UNIVERSITY OF CHICAGO PRESS
Chicago and London

Noel Kingsbury
is recognized internationally, through his books and journalism, as a leading innovator in horticulture and landscape. He is the author or coauthor of more than a dozen books on gardening, including Hybrid: The History and Science of Plant Breeding, *published by the University of Chicago Press.*

The University of Chicago Press, Chicago 60637
The University of Chicago Press, Ltd., London
 Published 2014.
Printed in China

23 22 21 20 19 18 17 16 15 14 1 2 3 4 5

ISBN-13: 978-0-226-43745-3 (paper)
ISBN-13: 978-0-226-11855-0 (e-book)
DOI: 10.7208/chicago/9780226118550.001.0001

Library of Congress Cataloging-in-Publication Data

Kingsbury, Noel, author.
Gardening with perennials : lessons from Chicago's Lurie Garden / Noel Kingsbury.
pages cm
Includes bibliographical references and index.
ISBN 978-0-226-43745-3 (pbk. : alk. paper) — ISBN 978-0-226-11855-0 (e-book) 1. Perennials. 2. Lurie Garden (Chicago, Ill.) I. Title.
SB434.K567 2014
635.9'32—dc23

2013022933

∞ This paper meets the requirements of ANSI/NISO Z39.48-1992 (Permanence of Paper).

Contents

Introduction, 1

1 **The Story of the Lurie Garden,** 5

2 **It's Either Freezing or Baking—Gardening in a Midwestern Climate,** 16

3 **Making the Garden a Better Place for Plants,** 29

4 **Choosing Perennials for the Garden,** 45

5 **Putting Plants Together,** 55

6 **The Gardening Year—A Guide to Essential Maintenance through the Months,** 67

7 **The Wild, the Native, and the Cultivated,** 75

8 **Plant Directory,** 86

Resources, 197

Acknowledgments, 199

Common Names and Scientific Names, 201

Index, 205

Introduction

Since its opening on July 16, 2004, the Lurie Garden in Chicago's Millennium Park has attracted an enormous amount of attention. Hundreds, sometimes thousands, of people pass through it every day, often on a regular basis. Its rich array of plant life creates endless talking points as it grows, flowers, and dies back through the year. The wild birds and insects that visit have become a big part of the garden experience too. Many of its visitors with gardens of their own have wondered about how they might have some of the Lurie magic back at home.

The idea for this book arose out of a conversation with Colleen Lockovitch, the chief horticulturalist at the garden from 2004 to 2009. We felt that there was a need to explain more about the plants in the garden to its many visitors and admirers, and in particular to explain how they could be used in home gardens. This book is the result—an introduction to growing perennials. It introduces some basic garden concepts, especially with regard to a continental climate, like that of the American Midwest. Colleen's successor as horticulturalist, Jennifer Davit, has also been involved with the book.

Perennials offer more for the home gardener than any other group of plants. They are small enough to make possible creative planting design in restricted spaces and long-lived enough to develop continuity and improve the sustainability of the garden. In writing this, we very much hope that more people will realize the pleasure, peace, and passion of gardening and working with plants.

This book introduces garden perennials, using the plants of the Lurie Garden as a guide. It is not intended to be a comprehensive guide to perennials but starts with the idea that the range of plants used in the Lurie Garden forms a good basis for

home gardeners new to growing perennials or indeed to the whole world of gardening. The range of perennials in the Lurie Garden is actually very wide, although with an emphasis on species that flourish in sun rather than shade. It also includes many regionally native species—more than half.

Readers may wonder how come an Englishman is writing a book about a garden in Chicago. They might also like to ask why the garden was designed by a Dutchman (Piet Oudolf). The answer is that gardening has become a very global business. Many of the plants American gardeners grow are of European or Asian origin, and many of our garden plants in Europe are North American. Seeing plants grow in different places is an important part of the job, and seeing them growing in the wild is a particularly special privilege. I shall never forget traveling in Pakistan in 2005 and seeing Russian sage (a Lurie Garden plant) and several other plants familiar from gardens, growing wild in incredibly harsh conditions.

Over the last few years, I have undertaken research on long-term plant performance and learned a lot about how plants live and spread—and sometimes die! Many of these factors are unique to the plant, no matter where it is growing. Choosing plants that will survive for many years, but also not spread invasively, is crucial to satisfying and sustainable gardening. The plants in the Lurie Garden have been chosen for reliability and longevity. The majority will thrive on a wide range of soils and across a wide climatic range. As we shall see, the plants here can thrive with minimal irrigation, no fertilizers, and no chemical control of pests and diseases. This is a very modern garden in its sustainability. The "Lurie flora" is wide enough to allow for considerable experimentation in the home garden, but is also a truly tried-and-tested range of plants.

The world of perennials is one that is changing rapidly. Many of the Lurie Garden plants are relatively unfamiliar and new to the nursery trade. Almost certainly they will become more familiar, as the garden is not only influencing garden designers in the region but the nursery trade too.

TOO MUCH LAWN?

Whenever I see all those acres of American lawns, I feel guilty! The lawn is a British idea, and in these more sustainability-conscious times even we are having second thoughts about it. Many gardeners, on both sides of

0.1. ***Looking over the Lurie Garden from the south, from the roof of the Art Institute of Chicago in late May with the Salvia River making an impact. The area of planting called the Light Plate is to the left of the diagonal path, and the Dark Plate is to the right. The surrounding hedge is the Shoulder Hedge. (Photo: Linda Oyama Bryan)***

the Atlantic, feel that we can live with rather less of a garden feature that needs irrigation in summer; regular feeding, pest control, and weeding to look its best; does not do anything for wildlife (apart from the odd bird pulling out a worm); and looks the same all year round (except when it's covered in snow or turned brown in the summer). Lawns are great for sunbathing, childhood games, or hosting barbecue parties, but not for much else. Why not reduce their size and grow perennials instead (or shrubs for that matter)? A selection of perennials offer interest at all seasons from spring to midwinter, can be managed more sustainably than a lawn, and offer far more for garden wildlife.

AMERICAN NATIVES AND GARDEN EXOTICS

A long-standing debate in gardening circles is over the role of native plants. Two issues have pushed this to the top of the agenda. One is that of invasive species, and the other is the role of garden plants in supporting wildlife. The design of the Lurie Garden avoided using any species that might become invasive, that is, spread beyond the garden and potentially into the wild, but there is nothing controversial about this. What is more controversial is just how important it is to use only native species. Ecologists have demonstrated how the web of life in natural environments is dependent on invertebrate species that will only feed on certain (native) plants. They argue that a garden only planted with exotics (i.e., nonnative species) will be an ecological desert. Many gardeners disagree, and, more fundamentally, point out that most people garden to first and foremost create a pleasant environment for themselves, their families, and friends; having attractive and easy-to-maintain plants are their priorities.

The style of planting in the Lurie Garden, and which this book is trying to promote, offers a possible resolution to the native/nonnative plant debate. Over half the species used are Midwest natives. Of those that are not native, many still have value for wildlife: bees are not fussy about whether the flowers they suck nectar from originate in China or the Midwest, and birds do not care whether the seed they are eating from perennial seed heads is of North American or European origin. One of the things that is really special about the Lurie Garden is how it brings together the native and nonnative, helping provide a sanctuary for urban wildlife and for human city dwellers.

In this book, we first look briefly at the story behind the Lurie Garden and its roots in a planting style that puts perennials first. We then look at some of the issues involved in gardening with perennials in a continental climate of hot summers and cold winters. In the chapter "Making the Garden a Better Place for Plants," we look at how garden conditions can be improved for growing perennials. I try to help with difficult decision making at the garden center in "Choosing Perennials for the Garden" and "Putting Plants Together." The chapter "The Gardening Year" is intended as a brief summary to help readers with notes on seasonal garden care. In "The Wild, the Native, and the Cultivated," I try to draw a link between perennials in the garden and the remnants of the wild prairie landscape in the Midwest; increasingly, as people grow more native plants and begin to manage their gardens for wildlife, we make greater connections with wild plant habitats and their conservation. Finally, the core of the book is a directory of the Lurie Garden's plants.

The Story of the Lurie Garden

Millennium Park was the final piece of the jigsaw puzzle for the development of the lakeside area of Chicago. Most of that area, which had been part of Daniel Burnham's 1909 plan for Chicago, became Grant Park. Millennium Park, 24.5 acres in all, was thought of as a covering for an underground parking garage, itself built over the tracks of the Metra/Illinois Central Railroad. The making of relatively enclosed gardens in public parks is an old tradition, a way of concentrating an ornamental horticultural element and providing a quiet place in the larger, and sometimes noisier, park environment. Richard Driehaus, one of the original founders of Millennium Park, offered to underwrite an invited competition for the design of a garden in Millennium Park. The Ann and Robert H. Lurie Foundation agreed to provide an endowment for the future maintenance of the garden, which was to be named the Lurie Garden, in honor of their $10 million gift.

After considerable jury deliberation, the competition was won by Gustafson Guthrie Nichol (GGN). The principal, Kathryn Gustafson, had a reputation as one of the world's leading landscape designers; originally a fashion designer, she turned to landscapes and made a name for herself with a number of projects in France (e.g., the L'Oreal factory and the Rights of Man Square, in Évry). The winning plan was developed in conjunction with theater and opera designer Robert Israel and Dutch garden designer Piet Oudolf. Gustafson and Oudolf were asked to submit designs separately, but given their complementary talents, they had decided that collaboration was the best way forward.

Oudolf's reputation had spread beyond the world of the garden with plantings for the Dreampark in the city of Enköping,

Sweden, and a number of visitor destination projects in England: for the Royal Horticultural Society garden at Wisley, near London, and for the Pensthorpe Waterfowl Trust in the east of the country.

Gustafson describes meeting Oudolf: "When I was considering the Chicago competition . . . we'd both read the brief at the same time . . . I thought his work was extraordinary, I picked up the phone and suggested we collaborate." Collaborations between professionals in landscape and garden design have been rare up to now, but in Gustafson's view, "I think Piet has changed the way landscape architects see gardeners and horticulture professionals, they see them with a better understanding of what they can bring to a project . . . we had decided that we don't know this type of perennial planting well enough and we need to bring in expertise." The Lurie Garden is a striking example of the success of such collaboration, but unlike great works of architecture, it has much to teach the ordinary home gardener, which is the message of this book.

The Design of the Lurie Garden

The whole 5-acre Lurie Garden site is treated as a work of art, with a central concept, the realization of which is heavily dependent on the 2.5 acres of Piet Oudolf's planting. A wide boardwalk was included, and a narrow waterway (the "Seam") divides the site into two distinct regions of planting: the "Dark Plate" and the "Light Plate." The Dark Plate is an area of open woodland, richly underplanted with shade-tolerant plants—symbolizing the wild landscape that existed before the arrival of white settlers. The Dark Plate concept is about lush, relatively dark-toned and coarse-textured vegetation with enough trees to cast some shade. Considerably larger, and set at a lower level, is the more open, expansive, and fine-textured area of the Light Plate.

"One of the most important aspects of design for me is creating a project that emerges from its place," says Gustafson of her approach to landscape projects—what she designs, must, she says, connect, with the history of the location. The Seam waterway, for example, is an evocation of the historic edge of Lake Michigan, whose natural boundary once lay on the edge of this site, but which nineteenth-century railroad development pushed farther into the lake. The wooden walkway adjacent to the Seam echoes the old wooden sidewalks that once lined the city's streets.

The Lurie Garden is separated from the rest of Millennium Park by a massive hedge, the evergreen conifer arborvitae (*Thuja occidentalis*) and two deciduous trees: European beech (*Fagus sylvatica*) and hornbeam (*Carpinus betulus*). This is partly functional, sheltering the garden and its users from wind, but its muscular and monumental character is also intended to echo the well-known poem by Carl Sandburg (published in 1916) that describes Chicago as the "city of big shoulders." Kathryn Gustafson describes how "Shannon Nichol and I had worked on this with Bob Israel—we had decided we wanted a secret garden . . . the problem was that after a concert in the Jay Pritzker Pavilion [designed by Frank Gehry; located just north of the garden] up to ten thousand people would pour out and make for the entrances to the parking garage; the Shoulder Hedge was our way of protecting the garden."

The "Shoulder Hedge" has another function; during the planning of the Lurie Garden, the architect Renzo Piano was designing an extension to the Art Institute of Chicago, so GGN and Israel decided to tilt the garden toward the institute so that, with the Shoulder Hedge behind it, it became, as Gustafson says, "very theatrical . . . we set the stage for Piet to work in."

The area called the Light Plate makes up the bulk of the garden; open and sunlit, it is a place in which you are far enough away from the buildings in downtown Chicago that they can be appreciated as a backdrop. The garden is large enough to make it feel that you are not really in the city, that there is almost a sense of looking back at its architecture, that the great wall of steel, concrete, and glass that is Chicago has already been left behind on a journey to somewhere else. It is actually a very good place to admire this architecture. The ground of the garden gently rolls, and in this way it reflects much of the landscape of the Midwest. Oudolf's planting captures something of the landscape too; it is very much a stylized representation of prairie habitat. This is achieved by its informality (at least in contrast to the kinds of landscape usually seen in public places) but, in particular, by the use of scattered ornamental grasses—these always evoke in the mind's eye of the onlooker the memory of wild open spaces.

The ecologist might look at the Lurie Garden and say that there is nothing natural in it, particularly given that much of the planting is in the clumps that gardeners have conventionally used for perennials and

1.1. ***The original planting plan for the Lurie, designed by Piet Oudolf. There have been changes, reflecting problems with some plants, such as rabbit damage, but the vast majority of it remains as in this plan. (Design: Piet Oudolf)***

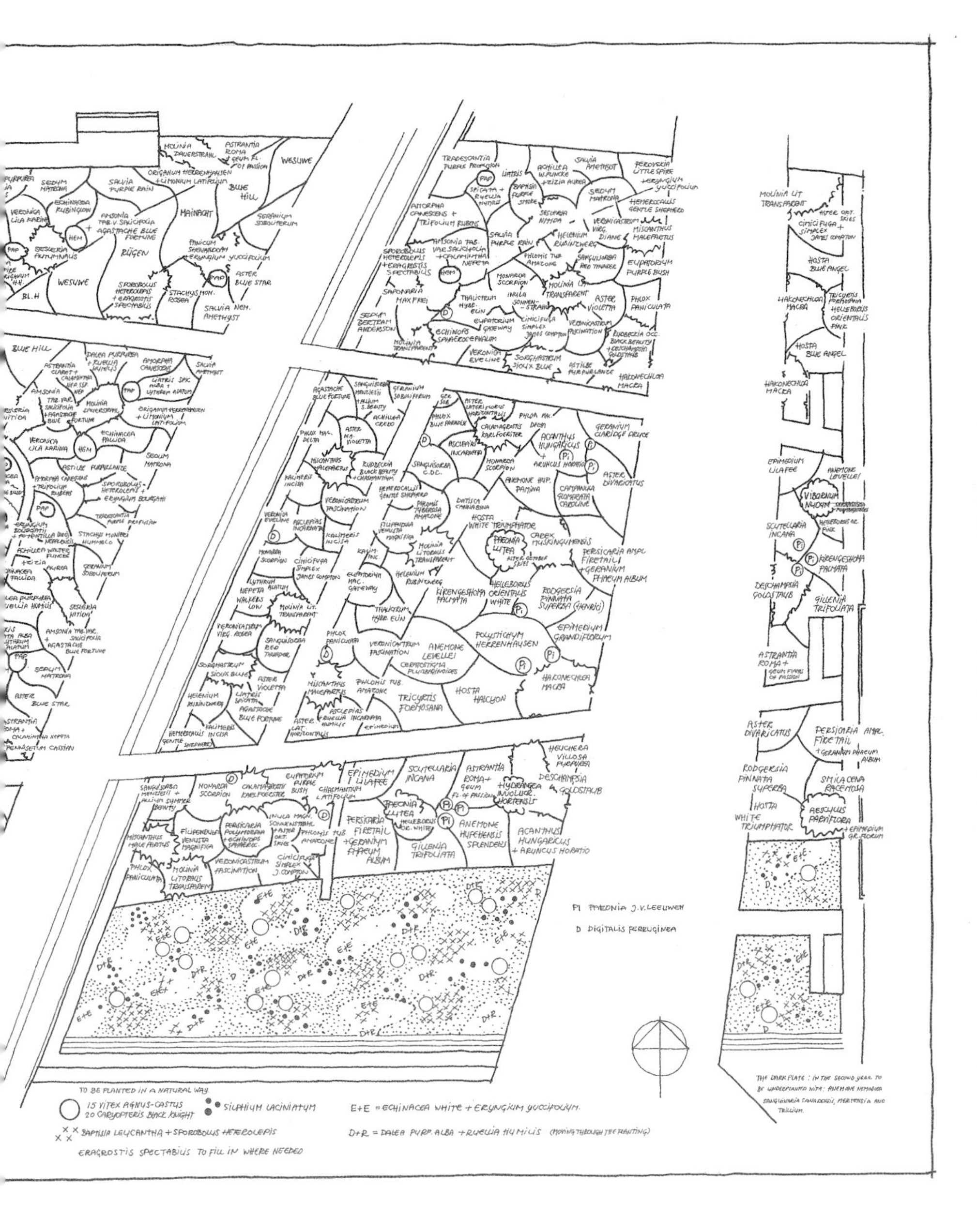

PI PAEONIA J.V.LEEUWEN
D DIGITALIS FERRUGINEA
THE DARK PLATE: IN THE SECOND YEAR TO BE UNDERPLANTED WITH: ANEMONE NEMOROSA, SANGUINARIA CANADENSIS, MERTENSIA AND TRILLIUM.
TO BE PLANTED IN A NATURAL WAY.
15 VITEX AGNUS-CASTUS
20 CARYOPTERIS BLACK KNIGHT
SILPHIUM LACINIATUM
E+E = ECHINACEA WHITE + ERYNGIUM YUCCIFOLIUM
BAPTISIA LEUCANTHA + SPOROBOLUS HETEROLEPIS
D+R = DALEA PURP. ALBA + RUELLIA HUMILIS (MOVING THROUGH THE PLANTING)
ERAGROSTIS SPECTABILIS TO FILL IN WHERE NEEDED

small shrubs. Yet, looking at the Art Institute Garden designed by Dan Kiley in 1965, it is possible to see how far landscape design has come; the earlier garden looks rigid and tame by comparison. It is also worth remembering that the Lurie Garden is in effect a roof garden, with a soil depth of between 18 inches and 4 feet of soil on a concrete base over a parking garage—truly "greening the city," creating habitat where there was none before.

Oudolf's work as a planting designer is always moving forward. He is forever experimenting with new plants, new ways of using them, and in particular new ways of combining them. The Lurie Garden represented a new level in the complexity and sophistication of his planting design; it repeated a number of elements that had proved successful elsewhere but also played with new ways of doing things. The bulk of the planting is in clumps, where multiples of individuals of the same plant species are put together. A few groups, however, are combinations, designed to set one plant off against another: *Geranium phaeum* 'Album' and *Persicaria amplexicaulis* 'Firedance', for example, flower at different times, the former early, the latter late, while *Amsonia tabernaemontana* var. *salicifolia* and *Agastache* 'Blue Fortune' put two blues together—making a harmonious combination of color but contrasting in form.

A key element in the Lurie Garden with great popular appeal is the "Salvia River." This was something that Piet first created in the Dreampark in Enköping, Sweden, using European species of *Salvia*, a group of plants closely related to culinary sage, with flowers in a range of violet blues; planted in a wave shape, and flowering at a time (early summer) before much of the Light Plate is in flower, the impact is electric.

More radical in design terms, and owing a lot to the naturalistic design movement (increasingly influential in both Europe and the United States), is intermingled planting, which aims to mix plant varieties more intimately than has been conventionally done. There are two ways of doing this in the Lurie Garden: one is to dot plants throughout to create a particular visual impact; the other is to create a more self-consciously naturalistic combination. This latter idea has been developed at the southern end of the garden (i.e., nearest the Art Institute), in the area Oudolf dubbed the "Meadow." This area has a matrix of ornamental grasses, which gives the feel of a much more naturalistic area; inserted into this are

1.2. *The Meadow is an area in the garden where plant species are intermingled rather than planted in clumps, creating a more naturalistic effect. The white is prairie native* Baptisia alba *var.* macrophylla, *the pink heads are the ornamental garlic* Allium christophii. *(Photo: Piet Oudolf)*

a number of perennial species that rise up above the grasses, the colors and textures of their foliage enhanced by contrast with them.

The Dark Plate, the narrower, higher part of the garden, is dominated by trees. Its character will develop more slowly over time as the trees mature, increasing the amount of shade at ground level. The perennials used here are all selected because they will be able to cope with decreasing light levels; they or their wildflower ancestors are species that originate from woodland or woodland edge habitats.

When it came to implementing the Gustafson and Oudolf plans, many people were, of course, involved. Two in particular were crucial: Terry Guen, whose company Terry Guen Associates was responsible for the implementation of the plans (the building, planting, and supervision of the establishment of the garden, and indeed of all of Millennium Park), and nurseryman Roy Diblik, whose partnership Northwind Perennial Farm grew many of the perennials (the other main supplier was Midwest Groundcovers). Guen described her role as "chief nurturer, problem solver, and rabbit patrol"—rodent damage being just one of the many problems that had to be dealt with in the construction and first year of the planting; indeed rodents have remained a problem. Roy Diblik's role began with working with Oudolf on plant selection during the planning stage. "[Oudolf] responded with such enthusiasm to prairie habitats," recalls Diblik. "He made a very emotional connection to places such as the Schulenberg Prairie at the Morton Arboretum] and took some of his impressions into the planning of the Lurie Garden."

Although he is a northern European, there is little doubt that Piet Oudolf has entered into the spirit of the American prairie in making the Lurie Garden. Working with local nurseries and plant experts, Oudolf was able to include a large number of native Midwest plants in his design for the Lurie Garden. In actuality, Europeans have a long history of using North American perennials in their gardens, and during the early twentieth century when perennial borders became a major component of garden style in Britain, Germany, and the Netherlands, a large proportion of the plants were American in origin. In falling in love with so many American plants, Oudolf is part of a long tradition.

1.3. ***Roy Diblik of Northwind Perennials and Terry Guen of Terry Guen Associates during installation of the planting. (Photo: Piet Oudolf)***

THE OUDOLF STYLE

The planting style that Piet Oudolf has evolved since the 1970s has its roots in northern Europe. Oudolf is Dutch; the Netherlands is a country with a rich horticultural tradition, both in plant production and in garden design. In using perennials, he drew very heavily on two sources: the work of German nurseryman, plant breeder, and writer Karl Foerster (1874–1970), who had popularized the use of grasses and ferns alongside conventional flowering perennials, and the British (or more accurately English) tradition of using massed ranks of perennials in borders. With his wife Anja, Oudolf established a nursery, primarily to supply plants for his design business, but which also proved very popular with private gardeners and, indeed, other designers. In seeking to stock the nursery, he undertook trips to British and German nurseries, looking for robust perennials that suited his style. Although for the first part of his career Oudolf had worked solely in private gardens, he was increasingly asked to design plantings for public spaces, in the Netherlands, Sweden, the United Kingdom, and the United States.

The Oudolf planting style has over the years changed quite noticeably,

1.4. ***Piet Oudolf in the Lurie Garden on a September day. Fall is already settling over the planting. (Photo: Rick Darke)***

but at the same time there are some core concepts that have not changed at all. The most important has to do with the selection of plants for structure: "I'm not a color gardener at all," he says. "I choose plants for their good structure, especially if it is long-lasting." Plants that are colorful and decorative but look a mess after flowering are not favored; those that may not actually be that colorful but which have good structure all through the growing season may well be. His saying that "a plant is only worth growing if it looks good when it's dead" may be a bit of an exaggeration, but it says something about a core component of his style—that plants that look good in seed should be as valued as those with flowers.

Finally, we must note that the planting at the Lurie Garden was taken one very colorful stage further, when, in fall 2004, some sixty thousand bulbs were planted, to provide interest before the perennials started performing. Many more have been added since. Bulbs mix with perennials very effectively, and in most cases the bulbs will continue to flower from year to year. A Dutch bulb expert, Jacqueline van der Kloet, was employed to select and supervise their planting. Jacqueline's particular expertise is in integrating bulbs with other garden plants and in evaluating them for long-term performance. Such longevity is an important part of the commitment to environmental and financial sustainability that is central to the mission of the Lurie Garden and vital to its role as an inspiration to home gardeners.

It's Either Freezing or Baking—Gardening in a Midwestern Climate

Gardeners tend to talk about the weather even more than anyone else. In the long run, however, it is climate that dominates everything that happens in the garden—year in and year out, through hot and cold and rain and drought. Understanding climate is crucial for the gardener who wants to make planting that works over the long term.

The American Midwest has a "continental" climate: cold winters and hot summers, with a rapid changeover from one to the other. Visitors joke that there are really only two seasons, with a week for spring and another week for fall. This clear pattern is more moderate along the North American East Coast, with winters less likely to be very cold and summers less hot (although perhaps more humid). In the Southeast, the winters are usually much less severe and shorter, but summers are infamously, unbearably hot and humid. The climate of the Pacific Northwest coast is very different and has more in common with that of northwest Europe, a maritime climate, dominated by the moderating influence of the sea, so winters are cool rather than cold and summers are hot for shorter periods. There is also a distinct Mediterranean element to the climate of the US West Coast—with a tendency for dry summers and wet winters.

In our gardens, we use plants from a wide range of different geographical regions. But our climates do limit what we can use. Gardeners in West Coast maritime climates can get away with using a surprising number of plants from much farther south, because their normal winters are relatively mild. A gardener in a continental climate cannot do this, as long periods of freezing temperatures and bitterly cold winds soon kill off plants that originate in climates where such weather is not regularly experienced.

The severity of the winter is the main limiting factor for what the gardener can grow in continental climates, although rapid and unpredictable alternation between warmth and cold can do a lot of damage to young growth in the spring too. In particular, winter cold is a major reason for the emphasis on perennials. Trees and shrubs are exposed to frost and wind, whereas perennials, which die down in the winter, are not affected by winter wind and the accompanying windchill factor. The fact that the living tissues of perennials are protected by being either at or just below ground level means that they are protected from the worst of the winter weather; snow too acts as a very good insulator. If you are a plant, being a perennial is a pretty sound strategy in a continental climate.

It should come as no surprise that the vast majority of perennials in the garden trade originate in continental climates. Of these, the overwhelming majority are from North America, mainland Europe, or central or eastern Asia. Bulbs are perennials too, but in a more compact form, and they too avoid the worst of winter by retreating underground; many also avoid the worst of the summer drought by going dormant by midsummer. Although the ancestors of garden bulbs are found across the northern temperate zone, the real center of diversity is central Asia, from Kazakhstan down to the mountains of Iran—an area of very marked continental climate and severe summer drought.

During the winter months, the gardener who works with perennials can relax. Everything just shuts down; they do not have to worry about winter gales or ice storms or deep snow—their plants are safely tucked out of harm's way. If, that is, they have paid attention to selecting plants that will survive in their hardiness zone. There is another advantage to gardening in a continental climate: because it is too cold for growth, there is no winter weed growth either.

Hardiness zones, developed by the US Department of Agriculture, are a very good guide to what is possible to grow in the Midwest and the eastern part of North America, but arguably less useful west of the Rockies. Although they have been used to describe minimum and maximum temperatures for plant survival and growth, it is really only the minimum that has been extensively researched and that is useful for gardeners. It

2.1. *The Salvia River is the high point of the Lurie Garden year and shows off the range of tones of blue, violet, and purple that are on offer from the versatile meadow salvias. These perennials are derived from a number of European species (*Salvia nemorosa, S. sylvestris, *and others) found in dry meadow habitats in central and eastern Europe, an area that has a continental climate approximately similar to that of the Midwest. Varieties of* Salvia *from left to right are 'May Night', 'Blue Hill', 'Wesuwe', and 'Rügen'; the mix then repeats. (Photo: Robin Carlson)*

is possible to use the Internet to enter a postal zip code and find out the zone for your area. Microclimate may affect the zone—being in a frost hollow or having a northerly aspect might tip a garden over into a colder zone, or having a southerly aspect or being next to a large lake might tip into it a warmer one.

The Lurie Garden itself is a good example of a favorable microclimate—being closer to the lake it is warmer than areas farther away, and the surrounding city reduces wind speeds and slightly raises the local temperature.

There is a strong argument that perennials should clearly be the dominant plants in gardens in continental climates, especially for those with small or urban gardens. But this does not mean that there is no role for trees and shrubs, whose scale and mass give them a vital role in the design of gardens. The beauty of particular species too ensures that they will always have a role to play.

ANNUALS, BIENNIALS, PERENNIALS— GROWING A WEB OF WORDS

It looks as if perennials are the group of plants that are the most useful for gardeners in continental climates. But what exactly are perennials?

A variety of words are used to describe plant growth forms. Unfortunately, there is a lot of confusion, partly because (as is so often the case when trying to describe nature) there are no hard-and-fast categories and partly because the words themselves are rather inaccurate.

Annuals live for a year. They have been the mainstay of conventional flower gardening. As short-lived plants, they have to produce seed to survive as a species, so they have evolved to have plentiful bright flowers, to ensure that pollinating insects such as bees notice them. Most annuals (and the most decorative ones) are to be found in desert or seasonally dry areas, where they exploit short periods of moist weather, produce masses of seed that can then lie buried until the next rains, which might be next year or in ten years time. The American Southwest and Mexico have given gardeners all over the world some of their favorite annuals, for example, California poppies (*Eschscholzia californica*), cosmos (*Cosmos* species), and zinnias (*Zinnia* species).

Biennials live for two years, flowering and producing the all-important seed to reproduce the species in their second year. The garden foxglove (*Digitalis purpurea*) is a good example. Biennials tend to be from unstable habitats; the foxglove is a woodland edge species, where tree growth can make an environment unsuitable for smaller plants in a matter of years.

Perennial is conventionally used to describe a plant that lives for multiple years, indeed potentially forever. In actual fact, many perennials are quite short-lived, so this term can cause confusion. Strictly speaking, trees, shrubs, vines, and bulbs are perennials too—which is why the term *herbaceous perennial* is sometimes used to describe those species whose growth dies down during a dormant period. *Herbaceous* means that a new set of shoots and leaves are produced every year from the same root; the death of the current year's above-ground growth and then dormancy is induced either by cold (in the case of relatively moist temperate climates) or drought (in seasonally dry climates).

Perennials are divided into two categories: *clonal* and *nonclonal*. *Nonclonal perennials* exist as a single rootstock, or possibly a very tight clump; they cannot spread except by seed. In some cases, these plants are very long-lived, for example, bowman's root (*Gillenia trifoliata*), but others are notably short-lived, such as coneflower (*Echinacea purpurea*). Short-lived perennials, like biennials, are often from unstable habitats where long-term survival is uncertain, so there is usually considerable investment in seed production.

Clonal perennials form clumps—they spread through producing new plants from their existing tissues, from roots (as with many perennial sunflowers, *Helianthus* species) or from stems that run along at soil surface (as with many cranesbills, *Geranium* species) or sometimes by shoots that "jump," as with cultivated strawberries. Each little bit with a shoot and a piece of root is potentially a new and independent plant—something that the nursery trade makes great use of when propagating plants through *division*, which simply means dividing up a clump of a perennial and making each bit of shoot plus root into a new plant. These perennials are referred to as "clonal" because they are capable of reproducing themselves clonally—each new bit is genetically identical to the parent. A brief inspection of the base of a clonal plant usually reveals shoots with emerging roots; if there are none visible, be aware that you may be looking at a short-lived nonclonal perennial.

Clonal perennials are generally long-lived because of their ability to clone themselves and spread. There is enormous variation in how quickly or extensively they are able to do this. Some plants only form very tight clumps with closely integrated shoots—an example would be the joe-pye weeds (e.g., *Eupatoriadelphus maculatus*, formerly *Eupatorium maculatum*), which are plants that can be divided, though a sharp knife and plenty of strength and patience are needed! Others send out running roots that can cover a huge amount of ground very quickly—"aggressive" may be a value-laden word to describe a plant, but it is often used for plants with this habit. Common milkweed (*Asclepias syriaca*) is a good example, a plant that even the nurseries selling it do not recommend for small gardens. Such plants are usually from rapidly changing environments or highly

2.2. ***Spring warmth brings bulbs out of the ground with remarkable speed. The Frank Gehry Music Pavilion seen over a drift of tulips ('Ballade' [pink with white edging] and 'Queen of the Night' [dark purple]) and daffodils (*Narcissus *'Thalia'). (Photo: Robin Carlson)***

fertile ones, like wetlands, where competition for space is intense. Many, indeed perhaps most, garden-worthy perennials occupy a middle position—an example is the goldenrod used in the Lurie Garden, *Solidago rugosa*, which, unlike some goldenrods, forms a neat clump, but which spreads out at a steady rate of 1 to 2 inches a year. Such a rate of spread is ideal for the home gardener—fast enough to look impressive in the year after planting, but not so fast that it makes the novice gardener nervous; wait a few years and there will be enough to divide, to hand out bits to neighbors, or pot up and sell in a charity garage sale.

Bulbs are basically perennials in a packet, a gift for the almost-instant gardener. They are usually from environments where there is only a short window for growth, such as forests or dry habitats. Growth needs to be fast, so having foliage and flower buds all ready to go, packed up in an underground container, makes sense. Technically, there are distinctions between *bulbs*, *corms*, and *tubers*, but for our purposes, we will call them all "bulbs."

BLINK AND YOU MISS IT—SPRING

Spring in continental climates happens quickly, as a rapid rise in temperature wakes up dormant buds and starts the expansion of stems and leaves. Temperature controls how long spring lasts—with continued high temperatures, flowers on spring-blooming plants will not last long, but if the rise in temperature is not so steep, bulbs and the ground-hugging perennials that are the main adornment of the springtime garden may continue to look good for several weeks. There is one great danger in spring, that of late frosts, when a sudden drop below freezing after several weeks of warm weather can do a great deal of damage to tender young growth. Whereas the impact on species that are not hardy, such as tomatoes or petunias, can be lethal, hardy perennials at least will recover.

There are a very large number of spring-flowering perennials that can be used in gardens, a great many of them being native woodland plants. In the Lurie, some of these are used in the Dark Plate. However, they tend to be slow to establish, and because they are slow to propagate, expensive for nurseries to grow—the wake robin (*Trillium grandiflorum*) is a good example. One that is used in the Dark Plate of the Lurie Garden is the Virginia bluebell (*Mertensia virginica*), an example of a woodlander that is quicker to establish.

In the Lurie Garden, bulbs are used to create the main spring impact. There are many advantages to using bulbs: they are relatively cheap, exuberantly cheerful, adaptable, and easy to grow. Many of the bulbs used in the Lurie are small species, originally from European woodland or mountain grassland habitats, such as *Anemone blanda*, and varieties of *Chionodoxa* (glory of the snow), *Crocus*, and *Muscari* (grape hyacinths)—flowering briefly in spring, with small leaves that die back in early summer. Most bulbs repeat flower well from year to year, building up into small clumps. Also included in the Lurie are some tulips, bulbs particularly appreciated for bringing color to gardens in spring and very early summer. Tulips, however, do not always flower well year after year; those in the Lurie garden are chosen from among those that do. The ancestors of garden tulips are from the Middle East and central Asia, where summers are hot and dry, conditions that, if not replicated in cultivation, may result in reduced flowering.

One of the very attractive special effects in the Lurie Garden, one that is easily reproduced in most home gardens, is to have small spring-flowering trees underplanted with bulbs. Two flowering cherries, the Sargent cherry (*Prunus sargentii*) and the Higan cherry (*P. subhirtella* 'Autumnalis'), and the redbud (*Cercis canadensis*) were used in the original planting. However, as Jennifer Davit points out, "these are not great trees for exposed areas in the Midwest, they prefer more of a woodland edge location. Fringe trees, *Chionanthus virginicus*, have been added to replace some of them." Spring may be short and sweet, but it is also a very busy time for the gardener. Planting needs to be carried out now. Spring planting is preferred to autumn planting in continental climates because when ground freezes and thaws during the winter, it can force newly planted plants out of the ground—a phenomenon known as "frost heave."

Perennial borders in spring are rather two-dimensional—as most interest is within a foot of the ground. Bulbs colorfully splash themselves around emerging perennials, most of which will appear as neat clumps. In some cases, these have a certain decorative value in themselves: the neat clumps of species of cranesbill (*Geranium*), the emerging tightly rolled leaves of hostas, or the divided leaves of burnets (*Sanguisorba*) are all interesting and attractive. Some leaves are colored red in the first few weeks after emergence, such as peonies and rodgersias; partnered with the blue of *Scilla* and *Chionodoxa*, they can make a striking impact.

HOT, HUMID, BUT A LOT OF PLANTS LOVE IT—SUMMER

Spring can turn to summer quickly, warm weather causing spring-flowering plants to go over faster than they might in more moderate climates. The last of the spring-flowering bulbs provide a link with the first summer perennials, in particular those dramatic ornamental onions/garlics known as the 'Drumstick' varieties (certain *Allium* species), and the camases—species of *Camassia*.

Perennials that flower relatively early in the summer tend to be short, but as the season advances taller species dominate. Midwestern prairie habitats are dominated by relatively late-flowering species, both grasses and flowering perennials. The Lurie Garden reflects this, with peak flowering from July to September. The role of grasses needs to be stressed here—in open natural habitats they tend to dominate, and yet their role in traditional gardens was minimal. The mix of grasses and flowering perennials that the Lurie Garden develops in summer can be seen as a way of echoing local habitats.

Summer in the Midwest is notoriously hot—typical for a continental climate—but it is humid too. We humans find humidity hard to deal with, and for plants it is a mixed blessing; high humidity means less evaporation, and so plants do not dry out so fast and growth can be rapid; but humidity does encourage many of the fungi and bacteria that cause plant diseases. Summer is not dry, however—in Chicago, August is, statistically speaking, the wettest month, with April to August being the wettest period of the year. It sounds like a great place to be a plant—so why do we humans water our gardens so much?

One reason we are rather obsessed with watering is historical, the legacy of a century of making lawns that, unless they get constant moisture, dry out and turn brown. The legendary green sod of the classic English lawn originally inspired Americans' obsession with making lawns and then trying to keep them perfect with constant mowing, feeding, and watering. But a dry summer in England means brown lawns too! Lawns brown off very easily, but—please note—do not actually die, as the roots are still alive and rain rapidly enables recovery.

Just because the lawn and, of course, plants in pots need watering does not mean that everything else in the garden does. The native perennials of the American Midwest clearly flourish in the local climate. So do perennials from a great many other places. So why do we feel the need to water them? Part of the reason is that in a dry summer, or a dry period in a "normal" summer, drought can cause leaves to begin dropping, growth to slow down, flowering to end prematurely—and, of course, we don't want a garden that is full of unhappy and stressed-looking plants. In nature, this is just part of the normal course of events, and unless the drought is extremely severe, plant communities recover when it rains again, or die down and start up into growth again next spring. But nature does not have to cater to human onlookers. An additional factor behind our desire to irrigate is that plants, perennials in particular, do undeniably look good

2.3. *During the deep, long Midwest winters the importance of perennials with remains (usually seed heads) strong enough to stick up through snow cannot be underestimated. This is* Astilbe chinensis *var.* taquetii *'Purpurlanze', a good example of a "three season" plant: it has good-looking bronze foliage in spring, deep-pink flowers in early summer, and the winter seed heads. (Photo: Robin Carlson)*

and healthy when generously watered, so there is an incentive not just to keep things alive, but to make them look as lush and luxuriant as possible. Plentiful watering can also keep some species flowering for longer too.

In places where there are shortages of water, it is important to take seriously all the different means by which water use can be reduced. Some of these involve soil modification or particular techniques (see page 32), but most crucial of all is choosing species that naturally cope well with low moisture levels, although ironically even drought-tolerant plants grow better and look better if liberally watered. Realistically, most gardeners will choose to do some irrigation. In hot and humid weather, however, irrigation can, having solved one problem (drought), contribute to another (disease). In particular, overhead watering helps create ideal conditions for a variety of bacterial and fungal disease problems. Bacterial leaf spot is a common example, where soggy brown spots, often surrounded by a yellow halo, appear on leaves. Overhead watering is in fact very wasteful as a lot of water does not get to the plants' roots (see page 71).

GOING OUT GRACEFULLY—FALL AND WINTER

Fall is usually celebrated by admiring the dramatic leaf colors that appear on many deciduous tree species. Perennials too can show good color, especially when a hot summer is followed by a sudden drop in temperature. Fall can happen slowly or quickly. Perhaps best for lovers of gardens is when there is a frost, followed by a succession of days, or even weeks, with clear skies, when daytime temperatures rise to a comfortable level and there are light frosts at night. Tender summer plants, like impatiens and dahlias, will be dead and brown, but hardy perennials can still look good and flower for weeks. Asters, black-eyed Susans (*Rudbeckia* species), and goldenrods (*Solidago* species) can flower for some time, while some perennials show good leaf colors, such as species of blue star (*Amsonia*), which turn a clear yellow.

The first serious freeze will kill all flowers, and mark the beginning of winter. But those who garden with hardy perennials can relax about fall, whereas those who rely on tender perennials and annuals are faced with frantically protecting plants or just saying good-bye to them; after a hard frost, all that will be left will be black mush, whereas the hardy perennial gardener still has the silhouettes of stems, seed heads, and often well-colored foliage to admire.

Traditionally, this was a time to clear away the dead remains of perennials. Now, however, many of us recognize the value of seed heads not just for their own stark beauty but also for the value that they have to winter wildlife, especially birds who feed on their hidden seeds and dormant insect life. Seed heads are potentially of great visual interest in fall and the early part of winter: they vary enormously in shape and color from plant to plant, some are even worth picking for use in winter floral arrangements.

Seed heads and the dead stems of perennials are very much a part of the plan for the Lurie Garden. Designer Piet Oudolf has always stressed the importance of their beauty, and his own photography of dead stems, some looking magical when covered in frost, helped popularize this look. It is an aspect of the garden that many onlookers found strange and "untidy" at first, but given the length of the winter in Chicago, it has been increasingly accepted as a worthwhile feature.

The seed heads of some will topple over in a strong wind, whereas others, grasses in particular, may stand firm for months, remaining as sharp silhouettes against the white of lighter snowfalls or glowing bronze in clear winter light. Given how bleak and empty winter can be, it makes sense to leave standing whatever there is in the garden in the hope that it may survive long enough to provide some pleasure.

Gardeners may chose to cut everything back and compost it before the first snows or wait until the weight of the snow flattens many perennials. Nurseryman Roy Diblik speaks poetically of the decorative impact of snow, describing how many perennials have "stems which are strong and hold up well to early snow . . . as the snow storms add up most of the foliage collapses, leaning and breaking stems in various directions, it can actually look quite nice, having interesting shapes and forms . . . by mid-February there are only a very few full structural plants standing." Jennifer is keen to stress that some plants will stand, recalling that "after the blizzard in 2011 we still had *Allium* 'Summer Beauty' standing, and *Monarda*, *Echinacea*, *Calamintha*, *Eryngium*, *Silphium*, and most of the grasses." "I'd encourage people to plant them to stand the snow," she says.

THE WIND NEVER BLOWS KINDLY

Gardeners tend never to like wind. Wherever it comes from, it seems to cause problems: a cold wind in spring can do a great deal of damage to young growth, while a dry wind in summer can seem to dry out plants quicker than the sun, and whenever the wind picks up strength its sheer force can break stems. Taller perennials are particularly vulnerable to having their stems broken. All perennials can suffer from rain damage, which is often exacerbated by wind; torrential rain causes lush foliage to become sodden and heavy, causing stems to bend and break. As stems grow taller through the growing season, such problems tend to increase.

Some perennials, though, seem to be almost immune to flopping and breaking. Anyone who has had to dig up joe-pye weed (*Eupatoriadelphus maculatus* and related species) or ironweed (*Vernonia* species) knows about their immensely strong stems and their radial root systems, which send out horizontal roots just a few inches below the soil surface—these are plants clearly designed to stand tall whatever the weather. Most perennials, however, are not made of such strong stuff.

Through the ages, gardeners have sought to design ways of supporting perennials, using stakes and string or metal support systems. A moment's reflection about natural environments, however, should raise questions—do wild prairie plants fall over? Exceptional wind or hail may flatten prairie wildflowers, but the answer is no: the plants have indeed evolved to stand up. In the garden, however, they are far more likely to flop or fall over. The reason is that garden perennials simply have life too good, as they usually grow more widely separated from each other than in nature, and so without competition they grow taller and are unsupported by neighbors. Cultivated varieties bred for flower size and plentiful flowers are often top-heavy. One more reason for perennials becoming top-heavy is that they are watered and fed better than in the wild, which makes for more luxuriant growth, but also structural weakness.

So, it looks as if the conditions of cultivation create their own set of problems. Gardeners can reduce the risk of perennial collapse if they do not feed plants growing on soil of average fertility; it is especially important to avoid using fertilizers that are high in nitrogen, which encourages plentiful but structurally weak growth. Plants in their first year are particularly liable to falling over, and many perennials will have a stronger root system to support them from the second year on.

Wind damage, either through physical harm or chilling or desiccation, will vary enormously from place to place. Even in one garden, there are places where a wind from a particular direction will do more damage than elsewhere: hilltops are notoriously windy, gaps between houses can funnel wind, and turbulence can develop as wind hits high walls. Perhaps more than anything else, wind exposure is something the gardener has to learn about over time: where prevailing winds come from, where they are strongest or most persistent, what time of year winds from a particular direction are likely to cause problems. Knowing where the wind is likely to blow from, it is possible for the gardener to set out new plantings in places where there is less likelihood of them getting damaged.

Making the Garden a Better Place for Plants

Choosing plants that are naturally suited to growing well in the environment in which they are planted seems obvious—but it was not always the case. The history of gardening lies in the production of fruit, vegetable, and other crops where the overall yield is regarded as of great importance; for our ancestors, survival depended on what they could grow. The key to increasing productivity is the idea that "the more you put in, the more you get out," so generous applications of water and nutrients (in the form of fertilizers or bulky organic matter like animal manure) were regarded as crucial. Applying inputs needed plentiful labor, and since lavish quantities of water and nutrients encourage weeds, labor would have been needed for removing these too. Ornamental gardening carried on in this tradition involved the cultivation of a relatively restricted range of plants, many of which also needed high inputs to perform well: roses that needed pruning, dahlias that needed to be dug up every winter and stored in frost-free conditions, phlox that had to be divided every three years, and so on. It is not surprising that gardening got a reputation for being hard work.

Traditional horticulture aimed at perfection, the growing of plants to a standard that was very human centered and—arguably—very artificial. Soils and situations that made the growing of the conventional range of plants difficult were defined as "problems." Throughout the twentieth century, however, there was a dissident faction in the garden world that instead promoted a more nature-centered approach: the use of native plants and informal, even wild, plantings and a partial rejection of conventional notions of tidiness. By the end of the century, this movement had gained ground, aided by the rising cost of labor to maintain the high levels of input needed for the perfectly striped lawns, neat rows of summer bedding, and precision-clipped shrubs of traditional gardening.

A contemporary and more ecologically informed style of planting and gardening does not try to change the environment to suit the plants, but to work the other way round—choosing the plants that suit the environment. Wild habitats are seen as sources of both inspiration and technical knowledge. A sand moraine, for example—one of the great heaps of material left by the retreating glaciers of the ice age—may be a challenging place for plants to grow, but there is still an interesting and attractive flora that will thrive there. The plants that flourish will be those that are naturally able to make the most of the limited reserves of moisture in the soil and to go safely into dormancy if it becomes too dry. A natural habitat like this teaches us that if we have a garden situation with a similarly dry soil, then we do not automatically have to think about extensive soil improvement or installing irrigation systems. Instead, we can use species that originate from a similar natural environment.

IT'S ALL IN THE GROUND—THE IMPORTANCE OF SOIL

Climate provides a set of constraints on what can be done, which, give or take the relatively small influence of local microclimatic conditions, are absolute. Soils also constrain what will survive and flourish, although they can be improved—given time and effort. Whether this time and effort is worthwhile is a decision the gardener has to make, but it is important to stress that, with good plant selection, it is not necessary to extensively modify soil in order to have a good garden.

Even nongardeners recognize the importance of soil, and the type of soil, to the business of growing plants. Soils vary enormously depending largely on the nature of the rock or other material from which they are derived. Very approximately, soils can be described as being on a gradient from "heavy" (dominated by clay) to "light" (dominated by sand). In the middle sits "loam," the perfect balance of clay, silt, and sand, which all gardeners would love to have, but few do. Ideas about "good" soil are heavily influenced by traditional gardening practices, and those of us who are mainly interested in growing perennials can be more relaxed and open-minded. The nature of soil does, however, accentuate certain climatic conditions.

Clay soils have a high proportion of minerals that tend to cling together and to hold onto water very tightly, making them "sticky." Digging a wet clay soil can be like digging glue, while digging a dry one is like trying to dig concrete. Prolonged wet weather or poor drainage results in long periods where the soil is sodden, whereas in dry weather soils can crack, which causes them to dry out more quickly, on top of which the soil minerals hang on to moisture so tightly that plant roots cannot access it. Many shrubs and perennials actually thrive on clay soils (roses in particular like clay); the problem with clay is physically working it, so a sustainable low-maintenance solution is to plant it with perennials and leave the soil mostly alone.

Sandy soils tend to be physically lighter, to hold together poorly, and to drain very quickly, with both water and nutrients running through them; hot dry weather will dry plants out on a sandy soil very quickly indeed. There are many drought-adapted perennials that will flourish, but in dry summers even they will become prematurely dormant.

Peaty soils are where there is a very high organic matter content but very little nutrient content. The underlying rock is usually acidic and low in nutrients; soil chemistry and high water levels so reduce the rate of decay of dead plant material that it builds up to form *peat*. Peat is used to make commercial potting composts because it holds its structure well, but it has to have considerable quantities of nutrients added to make it usable. In the garden, it is possible to add nutrients to peaty soils in order to grow perennials and other "conventional" garden plants, but it is more sustainable and more rewarding to grow what thrives naturally—rhododendrons, azaleas, and a host of other ornamental plants whose specialized chemistry enables them to cope with conditions that severely limit the growth of other plants. There are a great many North American natives that will flourish on these soils.

Modern approaches to choosing garden plants emphasize choosing species that will naturally flourish in particular soils; this is particularly the case with native plants. A particularly heavy or light soil should not necessarily be seen as a problem but as a guide as to what should be grown. In particular, it is now seen as important that we reduce the amount of water used for irrigation by paying more attention to plants that are naturally drought tolerant.

Having made the point that soils do not need to be the perfect loam of traditional practice, the reality is that sometimes the gardener is faced with some very poor quality soils. Often these are not the result of nature but of the construction industry or even some landscape contractors; a typical situation is where a housing development or construction site has resulted in topsoil being replaced with subsoil or where the soil has been compacted by heavy machinery. It is useful here to separate out the nature of the soil itself from any structural problems such as compaction.

Compaction may be obvious, or it may hidden—with a "hardpan" at depth that slows down drainage resulting in waterlogging that will also impede root growth. Problems such as these are best dealt with by landscape contractors who can use appropriate machinery to break up hardpans or open up compacted soils or even, in extreme cases, install underground drainage pipes.

Very poor quality soil, such as subsoil, is something that cannot be wished away, except by those wealthy enough to dig it all out, carry it away, and replace it with something better. The reality is that many gardeners do want to change their soil, particularly if having a very heavy or a very light soil exacerbates the effect of drought—the need for irrigation can be greatly reduced if soils are able to not only hold water but to release it to plants when they need it. For soils at both ends of the spectrum, the solution to soil improvement is the same—organic matter. The remains of life, in the form of manure or composted vegetable and plant matter, break down in soil to form humus, a chemically complex material that improves soil structure and, in particular, holds onto water and nutrients, but at a level that enables plants to access them easily. A limited amount of humus-building organic matter can be provided by a domestic compost heap, but many gardens with particularly problematic soils may need material brought in from outside.

Precisely what organic matter might be available locally will vary. In some areas, various agricultural waste materials are available, such as mushroom compost. Increasingly, composted green waste is available, the result of city-organized composting of domestic organic waste, au-

3.1. *The mysterious dusky gray purple of leadplant,* Amorpha canescens, *is not immediately outstanding—it tends to look best in combination with other colors, particularly blues and purples. It is a drought-tolerant native of the Midwest and a good example of the kind of less conventionally attractive plant that we need to start using to help improve the sustainability of our gardens. (Photo: Noel Kingsbury)*

tumn leaves, or shredded branches from tree pruning. Dug into the soil, or applied to the surface as a mulch, such material can do much to improve difficult soils, in particular to improve the water-holding qualities of dry soils.

Most soils fit onto the heavy-light gradient, but not all. Peaty soils are a special case and related to them, the very humus-rich soils that develop in mature forest. If the trees are felled and the soil exposed to the sun, then it will tend to dry out and break down; but if houses are built in an area of woodland, then gardeners may find themselves with just such a soil. Such soils should be cherished as they support some of the most distinctive, but also slow-growing, of native plants, woodlanders like May apple (*Podophyllum peltatum*) and the trilliums.

Finally, it is worth considering the issue of poor drainage. Soils can become waterlogged for a number of reasons, but the results are the same—waterlogging leading to the inability of roots to breathe, their death, and then infection by various fungal pathogens. The key issue is stagnation, as soils on slopes can become very wet, but because water is constantly moving, deoxygenation is less likely to happen. If a wet soil is the result of the lay of the land, it may be possible to improve drainage, but only after considerable work and expense in installing drainage pipes. An easier option is to concentrate on growing plants that naturally grow in wetland situations, such as mallows (*Hibiscus* species).

FERTILITY—TOO MUCH OF A GOOD THING?

We have seen above how the history of gardening was dominated by the need to increase soil fertility—this was vital for producing food and useful crops but not really necessary for ornamental plants. Nevertheless, feeding ornamental plants is what people did. Supplying nutrients to plants often involves importing material into the garden; in the nineteenth century when labor was cheap and every farm and many small businesses had horses, this meant generous applications of horse manure, which is high in nutrients, particularly nitrogen. During the twentieth century, the synthetic fertilizer industry revolutionized agriculture and, as a by-product, began to market their goods to domestic gardeners too. Any walk around a garden center will illustrate this—there are countless products that promise to make your garden flourish; whether they are industrially produced or homespun and organic, the idea is the same—you need our product to make your garden grow!

The fact is, however, that most of the plants we grow for garden ornament do not need, *and cannot even utilize*, the amount of nutrients we throw at them. Plants vary in what is technically called fertilizer response: some respond to increasing levels of nutrients by growing ever bigger, faster, and lusher, but many others may grow a little more with an application of fertilizer but then beyond a certain point will stop responding. A great many of the plants we grow in our gardens are originally from environments where nutrients are limited, and their physiologies are geared to survival with very little—be generous to them and they do not respond. In particular, plants from environments with very sandy or stony soil or from dry climates or from wetlands on poor acidic soils have evolved to cope with low, often very low, levels of nutrients. They are good news for gardeners with poor soils, but they are bad news for the fertilizer industry—their products are not needed.

Looking at the business of feeding plants from the point of view of sustainability, there is little doubt that a vast proportion of the nutrients given to garden plants over the years has been wasted, and trickled off down into the soil, reaching the water table, and eventually contributing to the pollution of rivers, lakes, and the sea.

Plants need a wide range of elements in order to grow well. Many of these are needed only in tiny quantities and are available in most soils—they are called *trace elements;* most gardeners do not need to worry too much about them. Then there are the "big three": nitrogen, phosphorus, and potassium (the last often called *potash*); these are needed in larger quantities and are more likely to be in low enough concentrations in the soil to cause problems to growers of commercial crops, hence the fertilizer industry. Nitrogen is needed for healthy leaf and stem growth, phosphorus for a wide range of biochemical functions that are fundamental to plant growth and potassium for flower and fruit production. Once plants have the amount of phosphorus and potassium they need, they do not respond to further applications, but some plants continue to respond well to increased levels of nitrogen. These are usually plants from naturally fertile soils; many of the larger perennials grown in gardens are from such environments: tallgrass prairie, river valleys, or the lower slopes of hills where plentiful water and nutrients are constantly flowing downward.

Plants from naturally fertile environments have evolved to respond well to nitrogen because they are in intense competition with each other for space; if a plant does not spread its leaves wide or reach for the skies, a neighbor will get there first and consequently shade it out. There is, however, a price to pay for such positive responses to increasing nitrogen—more nitrogen means softer and lusher growth, weaker stems, more top-heavy flower heads. In the wild, this may not matter, as plants

3.2. ***Most perennials do not need, and cannot even utilize, the levels of nutrients that many promoters of garden products try to encourage us to use. These late-summer-flowering prairie perennials, species of* Aster *and* Solidago *(goldenrod), do respond to increased nutrients, but often to the extent that, in garden conditions, they become top heavy and fall over. (Photo: Noel Kingsbury)***

often support each other; but in the garden, it is a problem, as perennials become unstable and fall over. The answer is labor-intensive and often visually intrusive staking. Such plants are also often more prone to wind damage as their leaves and stems are softer or structurally overloaded. It is worth noting that in the Lurie Garden, there are very few such plants; most are shorter, so they are physically more resilient. Their medium height is also an indication that they are not from high-fertility habitats. Such plants are often easier to manage, as staking or support is never necessary.

High levels of nitrogen in the soil have other problematic effects too.

"Weeds"—unwanted and aggressively spreading plant species are generally very responsive to nitrogen too—so high levels of fertility in the garden often benefit them rather than the plants that the gardener is trying to grow. In addition, the soft foliage encouraged by plentiful nitrogen is more prone to fungal and bacterial diseases.

All of these problems are exacerbated if, as well as there being plentiful nitrogen, there are also high soil moisture levels brought about by overgenerous levels of irrigation. The message is clear: being too generous to your garden plants can waste resources, create unnecessary work, promote disease, and cause physical harm!

If your soil is very poor, to the extent that plant growth seems seriously hampered, feed cautiously. Stable manure is a good source of nitrogen and has the advantage that it helps improve humus content and soil structure as well as feeding the soil, whereas more concentrated fertilizers only provide nutrients, and in the absence of humus to hold on to them, they may disappear quickly from the soil. Organically based fertilizers (e.g., fish meal, bonemeal, cottonseed meal, and processed sewage sludge) tend to last longer than the purer "chemical" ones, releasing nutrients more slowly. Soils with a very high mineral content (such as sandy or stony ones) are usually lacking in humus—so adding organic matter (see pages 32–33) at the same time as feeding will help to build a more balanced soil structure that will hold on to nutrients long term.

One additional thing needs to be said about soil and preparing soil for new planting—it is essential that all unwanted existing vegetation (i.e., weeds) be removed. Competition from weeds is often the biggest problem in establishing new plants.

Weeds will exist as the buried roots of perennial species and as seeds of both perennial and annual ones. Perennial weeds are often the biggest problem and are best dealt with either by treating with a biodegradable herbicide (such as those containing glyphosate) or by smothering the area for a year with old carpet or thick black plastic sheeting, which starves the plants of light. A patient year of an empty space to get rid of all weeds is far better than planting in an area not completely clear and having to battle for many future years with emerging weed shoots. Weed seeds will, however, continue to germinate and are often best dealt with by mulching, which suppresses their germination.

TO MULCH OR NOT TO MULCH?

In natural habitats, at least outside desert conditions, the soil surface is covered with a layer of dead plant material: primarily old leaves, stems, and so on. This can often only be appreciated if you get down on your hands and knees and tease the base of the plants apart with your fingers—this can be seen both in natural prairie and in agricultural pasture land. The dead plant material is constantly rotting away at the base and being replenished by new growth at the top. Much of the fully decayed material is ingested by worms and other invertebrates who then pass it out underground, where it joins the humus portion of the soil. Over time, humus itself is broken down by microorganisms into carbon dioxide and water, which with time will escape to the atmosphere and be photosynthesized by plants, and the cycle will start again.

In the conditions of cultivation, this cycle can break down if organic matter in the form of dead plants are not added to the soil. We have already seen how humus plays an important role in holding water and nutrients in the soil and making it available to plants. In the garden, organic matter and leaf litter can be recycled through cutting back plants, composting them, and returning the compost to the soil—which crudely replicates the natural process.

Mulching is often thought of as compost or organic matter that is scattered over the surface of the soil and having the capacity to rot down, although there are other, nonorganic materials that can also be used for mulching; they do not rot down and are therefore semipermanent, but help reduce drying out and weed germination. Gravel and other stony aggregates are the most common of the so-called *mineral mulches* and are often used for decorative purposes among relatively slow-growing plants.

Organic mulches will slowly decay and be taken down by invertebrates, thus stimulating a healthy invertebrate fauna. They also serve other functions—suppressing the germination of unwanted weed seedlings and shading the soil surface—thereby helping reduce moisture loss. During winter and late frosts in spring, they help insulate the soil and reduce frost heaving.

Key to mulching is that the material used is not dug in, but forms a layer protecting the soil surface. The most common mulch of all is the application of chipped wood to the soil surface during the winter—the result is an attractive-looking, tidy surface. However, like all good things, this can be overdone. Extensive mulching has now become a very common garden practice, and one that is arguably causing more problems than it is solving. It is frequently applied so thickly that is suppresses perennial growth in spring. It can also increase humidity around the base of plants and so encourage fungal diseases, exacerbating the effects of overirrigation (see below). In the Midwest, a lot of wood chip is actually imported from states in the South, such as Georgia, with the consequent transport miles—so there is nothing remotely sustainable about its use. Wood chip mulch can be useful, but it only needs to be applied an inch or so deep and not necessarily every year. Local material should be used to reduce "transport miles."

The best mulches are those that rot down quickly, so that the organic matter they contain can be rapidly incorporated into the soil's humus content by the natural life of the soil. One obvious source of such material is the dead remains of the previous summer's perennials—larger perennials (tallgrass prairie species in particular) and grasses produce a prodigious quantity of dead growth by the end of the year, the largest proportion of which is stem. Getting this stem material to rot down on a compost heap can be problematic, so if it can be shredded and applied to the surface of the soil as a mulch, so much the better—two problems are solved in one go: the material is disposed of and the soil benefits from the mulch. The policy now at the Lurie Garden is to do this, using a riding mower with a mulching blade. "The results," says Jennifer Davit, "look better than bare soil and all the nutrients are recycled . . . we now rarely buy in mulch to use on the plants."

IRRIGATION

Irrigation allows us the luxury of having a garden whose planting style may be profoundly at odds with the natural capacity of the local environment to support it. In some regions, water resources, limited and precious at the best of times, are becoming dangerously depleted. As a consequence, garden irrigation is increasingly being questioned, and in some drought-prone states, it is limited by law. Many in the gardening community would stress the moral responsibility of gardeners to reduce their water use.

We have already seen (page 25) that one obvious way in which gardeners can reduce their water usage is by using plants that do not need as much water as "conventional" garden plants, but also that home gardeners like to see healthy-looking plants and so often want to do some irrigation.

Needless to say, far more water is often applied to gardens than is needed. One reason is that many gardeners do not know how to water properly; a favorite technique is to get out a sprinkler for a short period at frequent intervals—a lot of water ends up on foliage where it evaporates and what gets to the soil only moistens the surface. Dampening the soil surface only encourages superficial root growth, whereas what is really wanted is roots penetrating deeper down, as there are usually reserves of moisture at depth. Another reason for water wastage is automatic irrigation, which comes on because a timer tells it to, not because the soil is dry or plants on the edge of stress.

Colleen Lockovitch describes the irrigation policy for the garden as being on an "as needed" basis. "Most people," she says, "overwater, and if they have automatic irrigation they flick the magic switch and think they can go away for months . . . we monitor constantly and we do targeted irrigation, spot watering, even hand watering sometimes, but only if plants need it." Jennifer Davit, the current garden director, has continued this policy, stressing that "we would never irrigate the whole garden at once, as a lot of species will not need it."

Efficient watering involves giving the soil a really thorough soak—a lot occasionally is far better than little and often. Many gardeners do not water until the surface of the soil begins to become seriously dry, which encourages deeper rooting as plants search for natural moisture in the cool depths of the soil. Thorough watering will then replenish the soil at depth. Efficient watering also means thinking about how water is delivered—pipes that release water at ground level will waste much less than sprinklers, getting the water to where it is needed and reducing the risk of foliage diseases. There are a number of "leaky pipe" or drip irrigation systems available, mostly developed for commercial vegetable growers but also made in kit form for domestic gardeners.

Getting watering right and realizing that perennials are actually a lot

3.3. *Plants vary in their response to drought, a response that often has roots in the habitat from which a garden plant's wild ancestors come. The pink* Echinacea *variety in the photo is a plant from prairie or savannah habitats where drought is occasional—its response is to go into premature dormancy. The pale heads are the ornamental garlic* Allium *'Summer Beauty', a bulb that has evolved to retreat into dormancy in midsummer, thus naturally avoiding the worst dry conditions. The sea lavender (*Limonium platyphyllum*) comes from particularly dry conditions, with leathery leaves that will survive all but the worst droughts and a mass of tiny papery flowers that continue to look decorative in the worst conditions. (Photo: Noel Kingsbury)*

tougher than we might think can do a lot to make gardeners confident in creating plantings that will survive and thrive long term.

PESTS, DISEASES, AND WEEDS

Problems with pests and diseases have conventionally been major concerns of gardeners, as can be judged by the fearsome array of chemical weaponry on sale in many garden and hardware stores. Gardening is changing, however, not just because of health and environmental concerns over the use of garden chemicals but because a contemporary, more naturalistic style of planting tends to be less reliant on perfection to look good. A wilder style of planting is able to hide the occasional plant that has been nibbled or infected, as well as a few weeds. Planting density is greater than in conventional borders, so there is less scope for weeds to be able to germinate and grow. A diversity of plants helps reduce pest and disease problems since it spreads the risk—the majority of pests and diseases being very specific about the species they affect.

How much control of garden problems the individual does is very much a personal decision. Weed control is certainly vital for a successful garden, but a more pragmatic approach should be taken with pests and diseases. Because pests and diseases tend to be specific to particular plants, there is always the opportunity to stop growing some plants and take up others—flexibility and openness to the realities of gardening in a not always friendly world is necessary.

In the Lurie Garden, chemical control of pests and diseases has never been used. "We just did not want to open that door," says Colleen. Instead, "we do a lot of monitoring . . . but a great advantage of perennials, is that you can cut them back if they get badly infested with something, we did that occasionally . . . an extreme case was *Stachys* 'Hummelo', which got a root fungus; we tried new plants, lost them, dug out soil and replaced both soil and plants, but that was exceptional." In many cases, particularly where fungal or bacterial diseases are the issue, there may be a big difference between varieties of the same plant, so if one variety gets problems, try another. Both Colleen and Jennifer describe some varieties of *Sedum* (stonecrop) dying one after another; 'Matrona', however, has done comparatively well.

Recent years have seen Jennifer starting to use beneficial insects,

3.4. *Colleen Lockovitch working on plant installation on the roof of a shed built for garden staff in 2009. A surprising number of plants thrive on roofs—obviously a difficult environment, where drought, shallow soils (and, therefore, a hot root run), and low nutrient levels put plants under considerable stress. Choosing plants from similarly stressful environments in nature, and occasional irrigation, can, however, result in successful plantings. (Photo: Robin Carlson)*

predator species that reduce the numbers of certain pests, such as a mite to eat spider mites, whose infestations can damage perennials in hot dry periods, and ladybugs to eat aphids. These are all available to domestic growers online.

As we have seen (page 37), weed control before a garden is established is vital. If problem weed species start to appear later on, hand pulling of individuals or hoeing off of seedlings is important at an early stage—both operations are much easier with a mulch than with bare soil, as roots come out really easily. If more persistent perennial weeds with running roots begin to take hold (perhaps because the soil was not totally clean at planting time), then spot treatment with a herbicide containing glyphosate can be done safely. It is worth pointing out that some garden plants can self-seed to become a problem; this has been the experience at the Lurie Garden with a number of species. However, this is very difficult to predict, as seeding rates vary greatly from one garden to another. An example from the Lurie is *Phlomis tuberosa* (tuberous Jerusalem sage), of which Colleen says, "I love it and would defend it, but we had to dead head it, before the seeds ripened."

Mammal pests, such as rabbits and deer, can be particularly destructive. Rabbits have been a curse for the Lurie Garden from the beginning and have made the cultivation of some otherwise very good species all but impossible. In domestic gardens, rabbits and other rodent pests can be largely excluded by fencing, which is not an option for many public spaces. Deer, however, are more difficult to keep out and are a major problem for many gardeners in suburban areas. They tend to cause more damage to shrubs than they do to perennials, however; they do tend to affect grasses, and there are many perennials that they will not touch; anything with silver/gray or hairy foliage for example.

Choosing Perennials for the Garden

SUN, SOIL, ASPECT—THE GARDEN ENVIRONMENT

Plants often succeed best where they are growing in conditions that approximate their natural environment. So, before getting too carried away buying plants at the garden center, it is vital to get a sense of what kind of habitat the garden offers to plants. Light (sun or shade), soil chemistry (fertility, acidity/alkalinity), and moisture needs are the three key factors that control how well plants will grow, but different plants need or cope with different levels of these factors—so it is these that are referred to in most garden plant reference books, and often too on the labels and other information material at garden centers and nurseries.

Light

Full sun is a kind of "default position" for garden plants—most of the species we want to grow for decorative effect grow best in good all-round light. Many will also flourish in situations where there is direct sunlight for three-quarters of the day. Given that the "natural" vegetation of much of the Midwest is open prairie, it is to be expected that many native species flourish in full sun; however, savannah would also have been an important presettlement habitat (see page 78)—many natives are also happy in the light shade typical of such sparsely wooded environments.

Light shade refers to situations where there is direct sunlight for half the day or where there is very light or dappled shade caused by trees with a thin or high leaf canopy.

Full shade is where there is little or no penetration of direct sunlight. Species from woodland habitat will thrive here—such as those used in the Dark Plate at the Lurie Garden; although, because the trees are still quite young, the Dark Plate is still growing its shade.

4.1. **Penstemon digitalis** ***'Husker Red' is usually grown in full sun, but the wild plant can be found in woodland edge habitats in nature, indicating that it will thrive in light shade in the garden too. The*** **Hosta** ***'Blue Angel' in the background will, however, be much happier in full shade—its leaves may scorch in daylong direct sunlight. (Photo: Jennifer Davit)***

The presence of shade can be complicated by additional factors. Shade caused by buildings can be accompanied by poor quality soils (caused by building operations or foundations) or a "rain-shadow" effect. The trees that bring shade to the ground at their base can also reduce moisture and nutrient in the soil around them as their roots efficiently extract both. In situations like this, the best way to grow anything more than a few very tough, and often rather dull, species, like Japanese spurge (*Pachysandra terminalis*) or sedges (*Carex* species), is to improve the soil with organic matter that can rot down and produce humus (see page 32). Natural, healthy woodland soils have a top layer rich in organic matter that holds both nutrients and moisture into which shade-tolerant perennials will root.

Latitude makes a big difference to the strength of the sun. The farther north you garden, the weaker the sun will be, which makes it easier to grow shade-tolerant plants in light shade or even full sun. The farther south, the more that plants from shaded habitats will scorch badly in the sun; in southern regions it may even be possible, perhaps even desirable, to grow certain sun-loving perennials in light shade. In addition, it may be that in the Southeast, high summer temperatures, exacerbated by high humidity, may make it difficult to grow perennials from more northern latitudes or from higher altitudes.

Soil Chemistry

Many gardeners are familiar with the fact that some plants, for example, azaleas, will only grow in acidic soil conditions. Other plants have preferences too, but most of the perennials grown in gardens are not that fussy—more important is the fertility level of the soil, in particular the amount of the three key nutrients for plant growth: nitrogen, phosphorus, and potash. Soil fertility can be measured through soil tests, but, in fact, it is relatively easy for even novice gardeners to get a sense of how fertile a soil is. If the soil is obviously sandy, very stony, or pale yellowish in color, it is probably a poor, or infertile, soil. Plant growth will be reduced, and leaves may be pale. Rich, or fertile, soils tend to be dark, and plant growth lush and vigorous.

It is a common assumption, much promoted by the garden industry, that fertility is automatically "a good thing." In fact, most ornamental plants do not need a great deal of fertility to perform well (see page 34). Most perennials will grow well enough on most soils to be found in gardens. There are, however, some perennials that naturally flourish on soils too poor to support many other species, and so these are particularly useful for gardeners on infertile soils, sandy ones especially. An example would purple prairie clover (*Dalea purpurea*), which flourishes on very sandy soils.

Moisture

The first thing that many of us learn about plants is that they need water. Soil moisture content does indeed have a major impact on plant growth; in particular, some plants have a greater need for water than others—or, another way of putting this, some plants survive drought much better than others. Plants that thrive in dry habitats do not actually *prefer* to be dry; they are just better at surviving, so in a garden with a well-drained soil with adequate moisture throughout the year it is possible to combine plants from moist habitats and dry ones. Where drought is a regular occurrence, only plants that flourish naturally in dry places will make good ornamental plants. Plants that survive drought may not look that good either during a dry spell, but they are more likely to carry on growing and flowering and, importantly, will recover quickly and are less likely to suffer long-term damage.

Plants that flourish in damp soils are often those that support lush growth, like *Hosta*. They tend to be very sensitive to drought. They do not necessarily flourish in conditions of waterlogging, however. Waterlogging, where water sits in or on the soil for days or weeks at a time, reduces the oxygen content of soil, making for very difficult growing conditions; as a general rule, only plants that have evolved special physiological mechanisms for survival in wetlands will flourish. This is why improving soil drainage on very wet soils can make such a difference to what you can grow.

Microclimate

Many different factors make up a garden's microclimate. It is difficult to give hard-and-fast rules, as different factors will combine in different ways, and there may be great differences from year to year.

Aspects will alter how much light and solar heat a garden gets—a south-facing garden will warm up in spring quickly but also dry out more quickly, whereas a north-facing one will be cool in spring and may suffer from reduced sunlight, so that tall perennials arch outward, weakening their stems.

Exposure to wind can be important, but with big differences from year to year. North winds in particular can severely chill a garden in spring or fall, while strong winds in summer can cause desiccation and break the stems of taller plants.

Given the complexities of microclimate, it may take several years to really come to grips with the potentials and the problems of a new garden, although talking to gardening neighbors is a good way to learn about many of the local issues in the early stages.

4.2. *A flowering cherry (*Prunus subhirtella *'Autumnalis') in early May, with tulips and daffodils beneath and beyond. Spring plantings can be very colorful, but at the same time there is a stark simplicity about them; as the growing season progresses, perennials add more layers of complexity and interest. Whereas bulbs can make use of early season light beneath deciduous trees, perennials that grow throughout the summer will suffer reduced light once the trees have grown leaves, so they need to have some shade tolerance. (Photo: Robin Carlson)*

Using Information

Reference books and websites provide plentiful basic information about plant requirements. Books and websites that are regional in their scope are particularly useful. However, there is frustratingly little in standard information sources about long-term plant performance, particularly of perennials. The staff of specialist nurseries, rather than garden centers, can be a very good source of information, particularly those that grow a lot of their own plants. Experienced gardeners are often a mine of information, especially about longer-term issues. Books and websites are discussed on page 197.

CHOOSING PLANTS TO SUIT YOUR TIME—MAINTENANCE

Some people have a lot of time in which to garden, others very little. Hiring someone else to look after the garden is not necessarily the answer to those with ample income but little time, as so often those who work for garden service companies have little gardening skill or plant knowledge. Key to minimizing maintenance is choosing plants that do not require much work. Maintenance itself is an elastic term—there is a difference between what needs to be done to stop a garden from deteriorating and what simply looks tidy. Tidiness itself is a very personal issue.

Garden shrubs generally need little maintenance as they are by nature long-lived. The annual trimming of them, often into shapes, is not necessary and is entirely a cultural choice, and one many gardeners abhor, as it stunts the natural shape and development of the plant. At the other extreme, annuals, one-year-life-span plants loved by many as the colorful heart of the summer-to-fall garden, are famously high maintenance, almost by definition.

Perennials present different problems, as some are very long-lived, others short-lived; some spread, others do not. All of these factors have an impact on maintenance. These are the important ones to think about for maintenance:

- Annual cutting back. All perennials need some winter tidying up. Some, such as cranesbill (*Geranium* species) and knotweed (*Persicaria* species), produce soft growth that is easy to tidy up and rots down quickly on the compost heap. Grasses, especially larger ones, and tall upright perennials like *Eupatoriadelphus* produce tough stems that need cutting into small pieces or shredding before they will compost—and then they rot down only slowly.
- Short life span. Perennials are not all perennial! Many have a life span of anything from three years upward. Short-lived perennials will need replacing. A popular example might be some varieties of coneflower (*Echinacea*).
- Spreading. Perennials that spread slowly might be welcome in a small garden, but those that spread faster are a good thing if there is plenty of space; however, once they spread over and outcompete neighboring plants, they will need thinning out. Examples might be the white wood aster (*Eurybia divaricata*) and Japanese anemone (*Anemone x hybrida*) in smaller gardens.
- Self-seeding. Some perennials produce fertile seed that results

in self-seeding, for example, tuberous Jerusalem sage (*Phlomis tuberosa*) (see page 148). If thousands of seedlings result, they will need weeding out.

It is possible to create a number of broad categories for garden perennials based on their life span and ability to spread:

1) Short-Lived Perennials

This can mean anything between three and ten years. These very often self-seed, but many gardeners simply buy new plants. Coneflowers (*Echinacea* species) are a good example.

2) Longer-Lived Nonspreading Perennials

Eventual life span is unknown, but long enough not to be a replacement problem. Plants stay in one place and do not spread. Most of the perennials in the Lurie Garden are in this category.

3) Very Long-Lived Perennials

Often slow to establish, which means that care needs to be taken that they do not get overwhelmed by neighboring plants or weeds in the first two years. Once established, they form large root systems and are very long-lived. Some like the indigoes (*Baptisa* species) and blue stars (*Amsonia* species) slowly form clumps, others like *Gentiana* do not.

4) Perennials with Discontinuous Spread

Species that spread rapidly with shoots that root into the ground as they go, but where the older growth does not live more than a year or two. They often tend to deteriorate in garden plantings, which is why none have been used in the Lurie Garden. Two notable groups are yarrow (*Achillea* species) and bergamot (*Monarda* species) hybrids, both very decorative midsummer flowers but needing constant replacement or propagation. Some varieties of these are notoriously short-lived, but as Jennifer Davit notes, "We have had *Monarda bradburyana* in the garden for years and not had any problems with it and have not had to replace it.. We also added *Achillea* 'Moonshine', and understand from Roy Diblik it should be relatively long-lived too."

4.3. *Two coneflowers, varieties of* Echinacea. *These plants, natives of the Midwest, have become very popular. However, they are not necessarily long-lived as garden plants, despite the "perennial" tag. The one on the left is a form of* Echinacea purpurea, *the main important ancestor for garden* Echinacea; *it can be expected to live for five years and sometimes longer. Plants derived from this species also self-seed well. The one on the right is 'Art's Pride', deliberately bred by crossing several different* Echinacea *species—it is a more "artificial" plant, produced for its attractive color. However, it is less reliable than older varieties, often surviving in the garden for less than five years. (Photo: Robin Carlson)*

4.4. *The silver flower heads of bigleaf mountain mint,* Pycnanthemum muticum. *This is an example of a long-lived perennial that steadily forms an ever-expanding clump. It is closely related to the better-known* Monarda *(bergamot or bee balm), which also spreads but which does not form such stable long-lived clumps. (Photo: Robin Carlson)*

5) Perennials with Spreading Ability

Species that continually spread outward forming large clumps, potentially living forever. These clumps may need dividing after a number of years if they are not to overwhelm other plants. These can be recognized even as young plants in the nursery by examining the base of the plant—if there are several shoots pointing outward with young roots emerging from them, they will almost certainly be one of these. Many familiar asters, for example, white wood aster (*Eurybia divaricata*), perform like this.

Clearly, gardeners who want to minimize maintenance will use lots of plants from groups 2, 3, and 5. However, since many in groups 1 and 4 are very decorative, those with more time will often like to grow these.

Research into grouping perennials is ongoing, but in the plant directory (chapter 8), any established information on life span, spread, and self-seeding is given.

DESIGN ASPECTS—CHOOSING PLANTS FOR VISUAL IMPACT

It is putting plants together for artistic effect that drives many people's passion for gardening. There are several different ways of approaching this; it helps to establish which has the most appeal or is most likely to work:

- Season of interest. Most gardeners want as long a season of interest as possible, and in small gardens this is particularly important. However, this often results in diluting the impact of the garden at any one time, as top performers and nonperformers share the same space. Those with plenty of space may consider plantings that are geared for one season. In some cases, the garden environment may suggest this—for example, a space under a tree makes a good place for a spring garden. Planting for a long season of interest involves making sure that there is a good distribution of plants that flower at different times, and since perennials dominate the summer and fall, bulbs and flowering shrubs are used for spring.
- Color. Planning a color theme is something that many gardeners with an artistic bent very much enjoy, and the use of color in the garden is something about which many books have been written. Plantings are possible that emphasize harmonies or contrasts. It is important to remember that foliage color can play a major role, especially since it usually has a longer season than flower color. The availability of colors can change through the season too: there are many yellows and blue-violet shades in spring and autumn, while reds tend to be late summer and autumn.
- Structure. Piet Oudolf, the Lurie Garden's designer, has always emphasized structure and rates plants particularly highly if they have good structure over a long season; for this reason, he values perennials with attractive and strong seed heads in fall and winter. Texture is related to structure—it describes the visual feel of plants: soft, rough, glossy, matt, smooth. A good test of structure is to take a photograph and then convert it into black and white—a planting with good structure will still look good, but many colorful plantings fail this test.

Many gardeners, however, do just buy whatever they like. In fact, since we all have particular likes and dislikes, very often it turns out that we are making subconscious decisions, so with time a theme becomes apparent. Helping to develop a theme will make a planting more coherent, which is what we will consider next.

Putting Plants Together

BASIC RULES—SCALE

Selecting plants that are all suitable for the habitat that your garden presents is a fundamental rule of plant selection. Beyond this, the really basic issue is that of size and scale. It helps to think about this as happening at three levels:

1. that of the garden as a whole: this covers the role of trees and shrubs and of entire perennial plantings;
2. that of the entire planting;
3. within the planting—it is useful to think of the smallest scale of planting as revolving around the concept of "planting companions"—see below.

The Garden as a Whole

Too many perennial plantings are unambitious in scale compared to the rest of the garden, which is often dominated by vast swathes of lawn. By all means, start small, but as your confidence as a gardener grows, think bigger—not just in terms of the extent of a planting, but also the use of taller perennials and grasses. It does not follow that small gardens need little plants—although in developing a sense of intimacy in smaller gardens, some designers argue that "small plants only make a small garden feel smaller."

Scale within the Planting

Here are some basic guidelines:

- Take care not to hide shorter plants behind taller bulky ones.
- Tall thin plants can be effective among shorter ones.
- Be aware that some plants that grow large are actually

5.1. *The vivid purple pink of betony,* Stachys officinalis, *makes a striking impact in midsummer. Here, it is nicely offset with clumps of prairie dropseed grass,* Sporobolus heterolepis *(right), and the hummocks of* Baptisia 'Purple Smoke' *(rear)—a plant grown for its indigo-blue flowers but which has a longer season of interest with its neat, almost shrublike clumps of foliage. (Photo: Robin Carlson)*

5.2 . *Gardeners should not just rely on flowers for interesting effects. The wiry leaves of drought-tolerant prairie dropseed grass,* Sporobolus heterolepis, *are of interest all through the growing season, while in late summer and fall, the fine seed heads make for a distinctive garden feature—they also have an unusual but attractive scent. (Photo: Noel Kingsbury)*

very airy, an example being *Molinia* grasses, where it is possible to see through their stems to whatever is growing behind.

- Conventional planting puts taller plants at the rear of a planting that is only seen from one side (the traditional border) or the center of a freestanding planting (as in a so-called island bed).
- If a "prairie" type feel is wanted, then perennials that grow tall by the end of the summer can be planted on either side of a path,

but the path does need to be wide enough to cope with plants flopping over in rain—3 feet is usually adequate as a minimum, but be prepared to cut back plants at the edges.

- Take care to plant so that plants have room to grow—see the entry for "Spread" in the plant directory (chapter 8).

STRUCTURE

Color is well covered by many gardening books, but structure is much less so—it doesn't make quite the same impact on the "new books" table at the bookstore. However, relying on structure in perennial plantings can often provide a longer season and more sense of "backbone" than simply relying on color. Here are some guidelines:

- The proportion of structure in perennial plantings starts off at nothing and grows through the year, with late summer to early winter offering the most variety. As a consequence, early in the year, shrubs have a role to play in avoiding what would otherwise be a completely flat garden of bulbs and early flowering but ground-hugging perennials.
- The most useful perennials are those that continue to provide structure after they have finished flowering, like the firm, upright stems of Culver's root (*Veroncastrum virginicum*) or the texture and bulk of the rounded mass of foliage of blue wild indigo (*Baptisia australis*).
- Plants that do not have strong structure are nevertheless valuable, particularly for color, earlier in the year. They may be thought of as "filler plants."
- Upright plants can look awkward and gappy together, so lower clump-forming filler plant are often vital.
- Piet Oudolf and many other designers use a rough proportion of 70 percent structural plants to 30 percent filler plants.
- Grasses are among the most useful structural plants:
    - they have a long season of interest, often for several months;
    - they are often the main source of interest in early to midwinter;
    - they are remarkably tolerant of a variety of growing

5.3. *A fall picture of* Echinacea *(coneflower) seed heads outlined against the cloudlike seed heads of* Sporobolus heterolepis *grass, with a blue haze of* Perovskia atriplicifolia *(Russian sage) behind. This effect, of combining dark and highly defined perennial seed heads against pale and wispy grasses, is a very striking and easy-to-achieve look for the late-fall and early winter garden. (Photo: Linda Oyama Bryan)*

conditions, with only shade and either extreme wet or extremely dry or poor soils reducing growth.

How can we think about structure? Just as we have words for different colors, it helps if we use words to describe some basic structural categories. Here are some examples of the categories that Piet Oudolf uses:

- Uprights. Grasses and perennials with strongly vertical stems or flower spikes. Flowers arranged in spires are particularly visually striking.
- Globes. Flower and seed heads sometimes tend toward spherical shapes.
- Buttons. Some flowers or flower heads form masses of small rounded shapes.
- Umbels. The flower heads of many perennials are flat or gently dome shaped.
- Daisies. The composite flower head of members of the daisy family tend to stand out strongly.
- Plumes. Some flower heads of perennials and grasses are plumelike and, in the case of many grasses, orientate themselves gracefully in the lee of the wind.
- Clouds. Flower heads where large numbers of tiny flowers form softly textured masses.
- Linearity. The leaves of grasses, irises, and day lilies (*Hemerocallis* species) are a distinct contrast to those of most perennials.

Combining different structural elements creates interest, although too much contrast can look jagged and restless—just how much contrast you can handle is, of course, a very personal decision.

PLANTING COMPANIONS

A good way to start thinking about how to put plants together is to think small-scale, in particular at small combinations—of three or four or five plants. The idea of "plant companions" is to concentrate on what looks good together. In the plant directory (chapter 8), every plant has a list of suggested "Planting Companions"—this is inevitably very subjective, but it is based on the following principles.

Basic Compatibility

As a very general rule, planting companions need to be of a comparable size. At least, that's a good way to get started. When you have more confidence, you can start thinking about putting plants with more disparate sizes together. The directory's suggestions also concentrate very much on flowers—so companions need to flower at the same time; but if foliage is brought into play too, then the possibilities get much wider.

Fundamentally, planting companions need to thrive in similar environmental conditions. In the garden, if the soil is reasonably good, there is good light, and there are no particular waterlogging or drought issues, then it is possible to mix and match plants from quite different environments—the plant directory (chapter 8) assumes this. If conditions are more extreme, then you have to be a lot more disciplined about what you put together.

What Looks Good Together

This is very subjective—it's about the gardener being an artist. There are some plant combinations that everyone agrees are electric. One of Colleen Lockovitch's favorites, for example, is eryngo (*Eryngium bourgatii*) foliage and prairie smoke (*Geum triflorum*) seed heads with the flowers of the ornamental garlic *Allium christophii*; for her, it's "like a Disney film." Most combinations work because certain basic rules of aesthetics are met—rules to do with *harmony* and *contrast.*

Harmony

Color harmonies put colors together that blend and soothe: like different shades of the same basic color or pastels (pinks and blues) or hot colors (reds and yellows). Some gardeners like to plant "single-color" borders; even if not actually implemented, thinking about a single-color border can be seen as a good starting point: make a list of plants in a particular color and then think about what might harmonize with them. A good example is in early fall, when many different asters can be put together, which all have various shades of blue violet, along with other blues such as cranesbill *Geranium* 'Jolly Bee'. The net can be widened to bring in pink flowers, which offer a different color range but are a harmonious blend

5.4. *Rattlesnake master (*Eryngium yuccifolium*) and* Perovskia atriplicifolia *make for a fine combination, based not on color but on contrasting forms—which is the key design tenet used by Piet Oudolf in putting together his plantings. The haze of the* Perovskia *contrasts with the neat balls of the* Eryngium. *(Photo: Robin Carlson)*

with blue violets. The next stage would be to ask whether a little contrast might add a spark—which a yellow flower would do.

Contrast

Contrasts can be very effective, but rather tiring to the eye if applied everywhere in the garden. An easy and effective example is the late summer look of blue violet and yellow, often seen in prairie or roadside habitats, composed of asters and ironweeds alongside yellow goldenrods and various other yellow daisy family species. This is a contrasting combination that actually works well because of the way the eye tends to perceive these colors as opposites, and they balance each other out. The artistically adventurous might like to mix colors that many people perceive as "clashing" such as bright orange and magenta pink.

Habitat Combinations

A habitat that limits one of the key factors for plant growth—light, shade, or nutrients—does tend to dictate what can be put together. Usually plants from the same habitat share certain characteristics, such as the fine leaves of many drought-tolerant species or the broad leaves of many shade-tolerant plants, so that there is a natural tendency for species from the same habitat to look good together.

ARRANGING PLANTS

One of everything may work in a small space, but anything larger requires multiple individuals of the same species to work well visually, as the mind appreciates coherence through repetition.

In the past, it was an almost universal practice in larger perennial plantings to plant in groups, usually in odd numbers. Recent practice, influenced by naturalistic approaches, has favored experimentation with intermingling multiple individuals of different species. The Lurie Garden represents a stage in this evolution (see page 10).

Group plantings have many advantages, but also disadvantages. The landscape design partnership of Oehme van Sweden has popularized the mass planting of certain perennials; Oehme van Sweden's swathes of the yellow daisy known as black-eyed Susan (*Rudbeckia fulgida*) in a number of public plantings in Washington, DC, have become justly famous. Single-variety blocks are also easy to maintain. However, when they are not in flower, they are dull or may be untidy, a great disadvantage in the smaller garden. If a deer browses through or a fungal disease strikes, then the results can be disastrous!

Intermingling varieties enables more plants to be shoehorned in together, thus providing more scope for exciting combinations and for interest across the seasons, as well as providing for every contingency—if that deer or disease does pass through, it will tend to be selective. However, intermingled combinations can be harder work to maintain, as plants will compete with each other, and over time the more vigorous will need to be divided to stop them smothering smaller companions.

One approach is to do what Piet Oudolf did at the Lurie Garden, which was to use small blocks of plants but with occasional "scatter plants" dis-

5.5. *Species of* Eryngium *are what are often called "architectural plants," species chosen primarily for imposing form or structure rather than color. Here, eryngo (*Eryngium bourgatii*) makes an effective combination with* Veronica longifolia *'Pink Damask', with distinctly different forms but harmonious colors. Here, the* Veronica *was obviously flourishing, but there have been problems with it in the garden—it dislikes hot summers. (Photo: Robin Carlson)*

tributed randomly throughout. These were individuals of varieties chosen to provide a thread: for example, oriental poppy (*Papaver orientale*) 'Scarlett O'Hara' is dotted through parts of the garden so that there is impact over a wide area.

Among the most effective long-term structure plants are grasses—they have a long season so they provide continuity as well as structure. In the Lurie, prairie dropseed (*Sporobolus heterolepis*) and switchgrass *Panicum virgatum* 'Shenandoah' are used, among others; these are relatively wide spreading, but in smaller gardens, smaller or narrower plants can be used—one of the most effective is the strongly vertical feather reed grass *Calamagrostis* 'Karl Foerster'.

THE ROLE OF SHRUBS, GRASSES, BULBS, AND ANNUALS THROUGH THE YEAR

Flowering perennials may be the bulk of modern garden plantings, but other plant forms have a role to play, especially outside the main summer flowering season.

Shrubs take up a lot of space, but their bulk can provide the basic form of a garden, especially a larger one, and during late winter, shrubs, and trees, may be the only things standing, and without them the garden can seem bare and empty. Here, let's look at the role of different plant forms through the year:

Spring—Flowering shrubs and bulbs are the main sources of color and interest. As the season advances, perennials become more and more important. Bulbs are tremendously useful as they can be inserted between perennial clumps; the two coexist very well—think of spring, dominated by bulbs, as a layer superimposed on perennial-dominated summer.

Early Summer—Many perennials are now at their best, but few have reached any height. Apart from roses, few shrubs are in flower from now on.

Late Summer—There is a tendency for there to be slight drop-off in the number of perennials in flower. A few grasses may be making an impact. Among bulbs, only lilies (*Lilium*) are of any impact. Annuals will be looking good—many gardeners shoehorn them into odd gaps for summer color.

Early Fall—Later-flowering perennials put on at least a month of top-level performance, with many grasses looking their best now. Annuals can also be important. Some shrubs and perennials will develop vibrant foliage colors as night temperatures drop.

Late Fall to Early Winter—As frosts kill perennial flowers, their seed heads can still be an attractive part of the garden. Grasses will now be the most visually important part of the garden.

Winter—The extent to which perennials will be flattened by snow will vary with the nature of the snow and the type of plant. Many Lurie Garden perennials tolerate snow well. Shrubs, comparatively snow resistant, will often play an important role in the garden, providing some visual mass.

The Gardening Year—A Guide to Essential Maintenance through the Months

SPRING

- Maintaining bulbs. Bulbs make a major impact now, but to maximize their chances of flowering again next year, their foliage must not be cut back, or tied up, or even overshadowed by other plants (particularly for tulips). Gardeners on poor soils might consider supplementary feeding for bulbs, tulips in particular—high potash feeds such as tomato fertilizer will help to ensure flowers next year.
- Propagating. Well-established perennials can be divided now. Older plants of some perennial species can form large clumps, which can be split, and the resulting divisions replanted elsewhere in the garden or given away. Division can be carried out simply by digging plants up and pulling them apart by hand, but in many cases this is too difficult for anyone with less than superstrength fingers! Two garden forks placed back to back and inserted into the middle of the clump can be very effectively used to tease and pull clumps apart. Avoid using a spade if possible as this will cut through shoots and roots indiscriminately. Small pieces are best planted in a nursery bed or potted, so as to keep an eye on them. Larger divisions can be treated as new plants and planted in final positions.
- Planting newly purchased plants. Plants that have sat in containers all winter at ambient temperatures can be planted as soon as the ground unfreezes, but anything that has been under cover and forced into early growth

6.1. *Spring means bulbs. These daffodils (*Narcissus *'Lemon Drops') and tulips (*Tulipa *'Ivory Floradale') illustrate the way in which bulbs and perennials can be combined with other plants, dotted around between perennial clumps. The perennials will grow and help conceal the slightly scruffy-looking bulb foliage as it dies away. (Photo: Robin Carlson)*

may be badly damaged by a late frost, even though it may be a hardy plant—so do not plant until after the last frost date for your area. Soil should be broken up around and below the planting site to encourage new roots to penetrate, and, if possible, a few roots teased out from the new plant and spread out in the planting hole. Soil should be gently firmed around the new plant and then watered, which will help settle the soil around the roots. If the soil is very dry, fill the hole with water before planting. Do not add fertilizer or compost—it is now thought that plants establish best if they have to fend for themselves in their first year.

6.2. ***Coneflowers (this one is*** **Echinacea pallida*****) are quintessential summer native wildflowers. A source of herbal medicine, as well as garden interest, wild species such as this are now being used by nurseries to produce an ever-widening range of garden hybrids. For many though, the elegant natural beauty of the reflexed petals of these flowers cannot be improved upon. (Photo: Robin Carlson)***

- Weeding. There may be some germination of weed seedlings as the weather warms up—these are most easily got rid of through using a hoe. This is also a good time to deal with more established weeds—if you suspect that emerging weed shoots are part of a larger root system, these can be sprayed with one of the herbicides (weed killers) designed to poison the root system of weeds, but which do not harm surrounding plants; the active ingredient glyphosate is the best known and environmentally safest of these.

- Mulching. Mulching (see page 38) is best done now, while the soil is still moist and cool. Mulching later onto dry soil may reduce the beneficial impact of rain as the mulch rather than the soil soaks it up. However, mulching can lead to disease problems in hot, humid weather, and it is best restricted to plants that are known to suffer in dry weather or where it is needed for the suppression of weed seedlings.

SUMMER

- Staking. Traditionally done to stop tall plants falling over. However, as we have seen (page 34), growing plants "lean and mean" can reduce this problem. Staking is very much a matter of personal preference and also depends on an assessment of the degree of damage likely to be done by wind (see page 27). If you do decide to stake, it is better to use one of the staking systems sold for supporting plants, and to get it up early—before it is actually needed. Trying to support plants that are already falling over rarely achieves a successful or tidy result.
- Pruning. "Deadheading"—the removal of dead flowers—is often done, at least by those with the time to do so; often it stimulates the production of fresh buds. An extension of this technique into what could be called "perennial pruning" has been developed by Tracy DiSabato Aust, an innovative horticulturalist in Ohio. She has discovered that light early or midsummer pruning can enhance or lengthen flowering of many perennials and can even be used to reduce the height of taller perennials—a useful technique for smaller gardens (see Resources). In the Lurie Garden, pruning of the sage (*Salvia* species) in the Salvia River helps encourage repeat flowering later in the summer.
- Watering. Whether and to what extent a gardener irrigates is a topic much argued over (see pages 24, 39). If watering is needed,

6.3. ***It is the leafy bracts that surround the flower heads of mountain mint (*****Pycnanthemum muticum*****) that make it an attractive summer flower, because these last much longer than the flowers themselves. Mountain mints are a group of native plants that have only recently become important in cultivation. (Photo: Noel Kingsbury)***

there are several ways in which the quantities used can be kept low and negative impacts (such as fungal infections) can be kept to a minimum.

- ○ Do as little overhead watering as possible. Instead, water at ground level, ensuring that roots get a good soaking. The best systems to install are those that rely on ground-level drip outlets around the bases of plants or porous pipe.
- ○ Watering should be infrequent and substantial.

- Pests and diseases. These tend to be a summer problem. Traditional gardening laid great store on minimizing them and a

6.4. **Calamagrostis x acutiflora** ***'Karl Foerster' is one of the most useful of all ornamental grasses. Shown here in late summer in Piet Oudolf's own garden, it has an immensely long season, from flowering in early summer to its seed heads, which stand well until late winter, resisting even hurricane-force winds. (Photo: Noel Kingsbury)***

"chemical warfare" approach. Naturalistic mixed plantings tend to reduce their visible effects because a wide variety of plants distracts attention away from any with problems. It is now well accepted that synthetic chemical controls are of limited impact and may cause more problems than they solve.

- Weeding. Odd weeds will continue to appear through the summer. In dense plantings, they will need to be hand pulled.

Where there are gaps between garden plants, a biodegradable herbicide can be used—see "Weeding" above.

FALL—WINTER

- Cutting back. At some stage, perennial plantings will need to be cut back to remove all the season's dead stem, leaf, and seed head growth. Seed heads can be very decorative and a good wildlife resource, so cutting them back is best delayed as long as possible. Apart from this, the decision about when to cut back is a very subjective one—how much untidiness can you stand? The easiest thing to do with cut material is to shred it and leave it on the soil surface to form a mulch that will rot down and recycle nutrients (it is possible to do this with a brush cutter, but takes some practice). Some, however, may find mulching too untidy and would rather pile their debris on a compost heap, from which material can be recycled onto the garden in a year's time. Jennifer Davit says that "I get a lot of questions from people about when to cut back their perennials if they leave them up throughout the winter. I always stress that they should cut everything back before the late winter or early spring bulbs emerge."

  Cutting back can be done in stages, so that untidy material is removed first, leaving the stronger and more visual stems (particularly grasses) for a few months more.
- Thinning out perennials. The more vigorous perennials will spread over time, which creates a risk that they will overwhelm less vigorous neighboring plants. Now is a good time to divide them, thus reducing their size, as the memory of where plants are will still be fresh (see Spring). However, replanting now may well result in frost heaving during the winter, so monitoring will be necessary when the garden thaws out next spring.
- Seed sowing. It may seem paradoxical to sow seed now, but many perennials have seed that needs chilling before it germinates (or, often, a period of warmth/moisture, followed by chilling). Finding out what pretreatment seed needs can be difficult, so often the easiest thing is to do what nature does and sow as soon

as the seed is ripe. Seed is best sown in trays of proprietary sterile compost and kept in a shady spot and regularly inspected once the weather warms up next spring.

- Preparing new areas for planting. Late fall and winter, when the ground is bare, is a good time to make major changes.

The Wild, the Native, and the Cultivated

WILD LANDSCAPES AND PLANTS

For the city dweller, the Lurie Garden hints at nature. The grasses and the gentle roll of the ground evoke the land of much of the Midwest. Traditionally, gardens were very much set apart from nature, but the Lurie Garden is an example of a more modern approach where nature is embraced and consciously evoked with a mix of native and nonnative species—midwestern natives comprise more than half the species here.

Here, we look at the connection between the plants used in the garden and the wild habitats from which they come, trying to make the connections with the home garden.

In the days before European settlement, the American Midwest was a patchwork of open and wooded land. In the East, trees would have dominated the landscape, as there would have been plentiful moisture to support their growth, but in the drier West, there would only have been enough water for more drought-tolerant grasses.

The prairie habitat itself is usually divided into "tallgrass" and "shortgrass," with moisture availability dividing the two. The division runs, very approximately, just west of the North Dakota–Minnesota state boundaries and then more or less straight southward. Tallgrass prairie was literally that, with tall grasses dominating the vegetation—big bluestem (*Andropogon gerardii*), the dominant grass in mature prairie, tops 6 feet. Mature prairie has less than 25 percent "forbs," that is, flowering perennials. The high point for perennial flowering is mid- to late summer; many are very decorative and quite a few have a long history of being used in gardens. Shortgrass prairie dominates not only in the drier West but in areas farther east where geology

7.1. *Early fall can be a very colorful time in prairie and savannah habitats. Here, several species of aster and golden rod (*Solidago*) grow among the dominant prairie grass big bluestem (*Andropogon geradii*) at a relict prairie area on the grounds of the Ragdale artists' community in Lake Forest, Illinois. (Photo: Noel Kingsbury)*

7.2. *A path mown through mature tallgrass prairie. The flower is* Heliopsis helianthoides, *one of the many daisy family species that dominate the mid- to late-summer period in this habitat. The trees are a reminder that the presettlement vegetation of the Chicago area was savannah rather than endless open prairie. (Photo: James Bodkin)*

leads to poor or dry soils. It is much less spectacular in visual terms, but not necessarily poorer in species, some of which have great potential for use as ornamental plants in dry areas, such as a key shortgrass plant, the grass prairie dropseed (*Sporobolus heterolepis*), used in the Lurie Garden and increasingly by garden designers in Europe.

Since the opening up of the prairie to European settlement and in particular to the use of the plough, the overwhelming majority of the tallgrass prairie has been lost to intensive agriculture—around 99.99 percent. This is a loss to nature, but it has undoubtedly done much to feed, not just the United States, but the world. Areas of relict prairie survive in protected lands such as national forests or because of some level of statutory protection. There are also strips of prairie along railroads. Degraded or partially developed prairie also exists in many odd patches or alongside roads. During the course of the twentieth century, however, interest grew in prairie restoration, both as a regionally appropriate landscape planting for the Midwest and as an ornamental and wildlife-friendly garden habitat.

Within such broad categories as "tallgrass prairie" and "savannah" were a host of very different habitats—their mix of plant and animal species dictated by a complex mix of factors, in particular, soil moisture and soil fertility. Prairie areas in river valleys often graded into wetlands, with vegetation getting steadily lusher as the soil got wetter. Dry sandy soils, particularly the heaps of sand left behind as glacial moraines, have a very different flora, characteristically sparse, but often including species found nowhere else. Such different habitats contain many wildflower species with considerable visual appeal for home gardeners, and, of course, many have evolved to survive difficult conditions.

PRAIRIE SURVIVAL AND RESTORATION

Areas of prairie survive, but in the Chicago area they tend to be particularly small relicts. Some are protected as nature reserves, while others are still vulnerable to development. Strips of land along railroads have been particularly important as refuges for plant species eliminated elsewhere, and it was from habitats like this that prairie restorationists in the twentieth century gathered seed so that they could start to recreate the habitat.

Areas of wet prairie and associated wetland sometimes survived if drainage for agriculture or other development was too expensive or impractical. Now, however, they are often seriously degraded as habitats because of what is known as "eutrophication"—nutrients (chiefly nitrogen but also phosphorus) in water from sewage or runoff from farms upset the natural balance. Plants vary greatly in their ability to utilize nutrients, and those that do so most effectively end up smothering everything else—cattail (*Typha latifolia*) in particular or invasive European forms of the reed *Phragmites australis*. Dry prairies may have avoided development for farming, but they have also been mined for sand or gravel or have been bulldozed for industrial facilities—many of the best sites have been lost.

Prairie is now recognized as a rich diversity of different habitats, each one of which itself contains and supports a diversity of plants and animal life. Formerly disregarded as waste land, prairie is now seen as a habitat of great importance and one that gives enormous pleasure to many people—prairie habitats are an integral part of many amenity and recreation areas. A number of citizen groups are now actively involved in prairie restoration projects. There is also a recognition of the value of prairie as an easy-to-maintain and visually appropriate landscape planting for the Midwest, and indeed beyond. Commercial sources of prairie seed and plants are now well established for many areas, with some nurseries now offering regional seed mixes.

Prairie conservation and restoration can involve

Clearing scrub and nonnative invasive species.

Saving seed from locally native species and sowing it in cleared areas.

Organizing burns—the best and most appropriate way of shifting the ecological balance away from invasive species to native prairie species.

Diverting or treating polluted water so that the growth of the most nutrient-hungry species is limited, allowing others to thrive.

The irony of the destruction of the prairie has been that many of its wildflower species have been important as garden plants; from the end of the eighteenth century onward, certain perennials from North America were imported to Europe where their ornamental character was much appreciated. At the turn of the nineteenth and twentieth centuries, the use of perennials in European gardens became highly fashionable, with

7.3. *Giant hyssop (*Agastache *'Blue Fortune') attracts a monarch butterfly. Butterflies are particularly attracted to flower heads where there are large numbers of very small flowers. (Photo: Robin Carlson)*

a key role for many late-summer-flowering North American species: asters, goldenrods (*Solidago*), and sneezeweeds (*Helenium*) being prominent and bred into showy hybrids. One hundred years later, the turn of the twentieth into the twenty-first century was a time when nurseries, garden designers, and others again turned to North American wild spaces in their search for new garden-worthy plants.

THE GARDEN, NATIVE PLANTS, AND WILDLIFE

Once upon a time, the idea of attracting wildlife into the garden would have seemed to go against the very idea of the garden. Nowadays, however, we are more likely to see the garden as an oasis for nature, a private nature reserve, or a place to observe nature. Fundamental to the idea of the garden as wild space has been the role of native plants; the native plant movement has been a vibrant but at times controversial part of the American gardening scene.

It has been forcefully stated that gardeners should grow more, or even exclusively, native plants. Among the arguments are that

- Only native plants support extensive and varied populations of insects and other invertebrates, and it is upon these that the more prominent, large, and often more "popular" wildlife, primarily birds, feeds and flourishes. In particular, many insect larvae will only feed on a particular plant species.

- Regionally native plants celebrate regional diversity and give an area a stamp of authenticity, as opposed to the bland uniformity imposed by garden center plants available everywhere within the same climate zone.
- Nonnatives have sometimes escaped and become invasive aliens, sometimes many decades after their initial introduction.

Many gardeners would argue to the contrary:

- The scientific evidence to support the idea that growing natives actually attracts more wildlife is often contradictory.
- While many insect larvae are very specific about food sources, adult insects and other animals are not—birds, for example, are happy to feast on berries whatever the country of origin.
- Gardeners want to enjoy a wide range of plants, and natives rarely supply enough year-round interest.

Much depends on how plants are organized in the garden. Some gardeners will want to sow a prairie or plant a mini-savannah in the backyard—and indeed they can be very attractive garden features. Most, however, will want more conventional lawns and borders. Native plants can play an exclusive, a major, or a minor role in such gardens. Native species that self-seed at moderate levels are a particular pleasure, as they are "doing what comes naturally" and help create a sense that the garden is in a continuum with the natural world of the region.

Native plants might be used in a garden in a way that is relatively formal and conventional, with stretches of bare earth between carefully spaced plants. Grown like this, they might not actually add much to local biodiversity. More fundamental for improving biodiversity than whether or not native plants are used in the garden is the way the garden is organized. Wildlife thrives in diverse habitats and in habitats that link up. In the Lurie Garden, a mix of natives and nonnatives (about half and half) and a rich diversity of plant forms and combinations provide a model for home gardeners.

Here are some pointers to the wildlife-friendly garden:

- Trees, shrubs, annuals, and perennials of various sizes are all key to a diverse garden.
- Cover and connectivity are important. Imagine yourself an insect trying to get from ground level into the branches of a tree; you want to be concealed from predators for as much of the way as possible, so intermediate layers of perennials, climbers, and shrubs will be useful.
- It helps to think of the garden in layers: beneath trees there can be shrubs and beneath shrubs there can be perennials. In more open borders with only grasses and perennials, there will not be layers as such, but there can be close connections between upright, clump-forming, and ground-cover perennials. Think of how closely mingled wild plants are in prairie and other natural grassland habitats and how far apart plants are in many conventional garden or park settings—closer integration is actually more "natural" and certainly more useful for wildlife.
- Do not be tidy. A little fallen debris and leaf litter on the ground or between established plants rarely detracts from a garden's appearance. At ground level, if mulch is not used, it is possible in established borders of perennials to see small "weeds" becoming established on the ground, alongside moss. These can play an important role for invertebrates and do nothing to detract from the decorative value of the planting.

Interest in biodiversity has resulted in it becoming increasingly common for the nursery trade to label particular plants as "butterfly friendly" or as "attracting humming birds." In fact, more or less any garden plants can be of some use to wildlife, especially if grown in the relatively integrated way described above. It is useful though to recognize some pointers for what makes particular plants food sources:

- Trees and shrubs with berries are important for feeding birds in the winter.
- Perennials with prominent seed heads that stand the winter well

7.4. *Dry prairie at Shoe Factory Road, very near Chicago's O'Hare Airport, in August. This plant community has developed on a sandy moraine; it is noticeably lower in height and sparser than prairie developed on more fertile or moister soils. This particular site also has more "typical" tallgrass prairie at a lower elevation—taller growth and more dominated by grasses. The pale-mauve flower is* Physostegia virginiana. *(Photo: Noel Kingsbury)*

attract birds, who feed not just on the seed but also on insect larvae that may be hidden inside.

- Perennials with masses of very small flowers in heads (such as many members of the daisy family) and which flower in mid- to late summer tend to be good butterfly plants. Examples include *Echinacea* species (coneflowers) and *Solidago* species (goldenrods).
- Bees also appreciate flowers with multiple heads. There are now beehives on rooftops in Michigan Avenue; the honey from these hives tastes slightly of mint because of the number of species in the Lurie Garden in the mint family, such as species of calamint

(*Calamintha nepeta*—not native) and clustered mountain mint (*Pycnanthemum muticum*—native).

- Perennials with tubular flowers have often evolved to be pollinated by hummingbirds, or at least are very attractive to them. An example from the Lurie is *Monarda braduriana*, the eastern bee balm, which is popular with both hummingbirds and bees.
- Double flowers are very rare in nature, and are rarely much use to bees, butterflies, and other insects. It is worth noting that there are no double flowers in the Lurie Garden.

The rewards of wildlife watching in the garden can be great, an extra dimension to gardening. In urban areas especially, gardens are an oasis for wildlife. The Lurie Garden is a good place to appreciate this, with migrating birds landing to break their journey from north to south or bees and butterflies finding food and places to breed. Organized wildlife walks have even become part of the Lurie Garden experience. With a little care, the wildlife experience is one that any gardener can have too.

Plant Directory

EXPLANATION OF CATEGORIES

Basic information about the perennials, ornamental grasses, and bulbs used in the Lurie Garden is given in this chapter. Some of the gardening concepts used in this directory are explained below.

Height

This is the *maximum* likely to be achieved, under garden conditions, of the plant when in flower. There are a number of provisos:

- Plants are often shorter, especially in poorer or drier soils or if grown where there is more competition; for example, if they are grown alongside other plants in a naturalistic planting such as a prairie, they may be very much shorter.
- Perennials with tall flower stems may have a mound of foliage at a lower height.

Spread

This is difficult to define, which is why many garden reference books do not attempt to provide this information. The measurement given here is what a plant is likely to achieve in three years after planting; it is also a good measure of spacing for conventional garden planting—imagine the plant to be in the center of a circle whose diameter is the measure of spread. The provisos here include the following:

- Plants can be grown closer together—after all, in nature, plants grow very close together compared to garden conditions; in more naturalistic plantings, this may be desirable, but there will be a tendency for plants to lose their distinctive shape and to become much more intertwined.

- Some perennials (see page 20) will continue to spread—potentially infinitely! For these, I have indicated how quickly they are likely to spread.

Season

This is given for flower interest in the Chicago area; farther south, flower interest could be earlier: farther north, later.

Garden Habitat and Cultivation

Most perennials are remarkably tolerant of a wide range of conditions; what is outlined here are the "outer parameters." The conditions needed may well be affected by the region in which the plants are grown—notably, the farther north, the more likely a shade lover will tolerate some sun; the farther south, the more plants suitable for "light shade" will actually need full shade. "Any reasonable soil" means a soil that is not excessively dry or wet, having any texture between unimproved clay or very sandy and at least average fertility.

Hardiness Zone—Minimum

The minimum is given, as cold tolerance has been well researched. Maximum zones are not suggested—these have not been not so well studied, and in my opinion many maximum zones given in garden reference books are misleading for gardeners outside continental climate zones such as the Midwest.

Planting Companions

For an explanation of what criteria I have used, see page 61. In order to limit what could be extensive lists, only companions among other Lurie Garden plants are chosen.

Scientific Names

Also known as "Latin names," these may look intimidating at first, but many gardeners use this botanical system as it reduces the confusion of common names and is internationally accepted. To explain very simply: a naturally occurring plant species will have a two-part name, written in italics, such as *Ageratina altissima*, indicating that this species is a member of the genus *Ageratina*, a group of plants in the daisy family (families being the next layer up in the botanical hierarchy). Selections made for their garden or landscape value and carefully propagated by nurseries to maintain a closely defined set of characteristics are often given an additional cultivar name, indicated by not being Latinized and written in regular type, such as the aster *Symphyotrichum novae-angliae* 'Violetta'. Hybrids are crosses between distinct populations; hybrids between different and separate species are given their genus name and an English-looking second name to indicate their artificial status; an example is *Aruncus* 'Horatio'.

Scientific names are notorious with gardeners for the way that some are occasionally changed. The most recent round of changes, which are shown here, and which particularly afflict the daisy family, are likely to be permanent, as they are based on the new and objective technique of DNA analysis.

FLOWERING PERENNIALS

Aconongon 'Johanneswolke' (*Persicaria polymorpha*) white dragon knotweed

Origin: A hybrid that appeared in a German nursery between two plants of Asian origin. The correct name, *Aconogonon* or *Persicaria*, is still disputed by botanists.

Height: 80 inches.

Spread: 60 inches.

Description: A gentle giant—at first sight this looks like a shrub but in fact is a perennial, collapsing with the first frosts to become invisible underground over the winter. In May, it reemerges, growing at a rate of an inch and a half a day! Do not be afraid—as, unlike its notorious relative the Japanese knotweed (*Fallopia japonica*), it does not run. Masses of creamy flower heads, turning pink with age, and a slightly disgusting scent. Very good value for almost instant impact.

Season: June, occasional flower later in the season, until September.

Garden Habitat and Cultivation: Sun or light shade. Fertile, moist soil preferable, but surprisingly drought tolerant once established. Needs

8.1. **Aconongon *'Johanneswolke'* (Persicaria polymorpha)** ***is easily mistaken for a shrub but, in fact, is a perennial, dying back to the base every winter. Its ability to create bulk quickly makes it useful for young gardens. Flowers emerge creamy white in early to midsummer, aging to dusky pink. (Photo: Noel Kingsbury)***

protection from wind. Forms a very tight clump, but roots liable to regenerate if the plant is ever dug up. Jennifer Davit adds that in the Lurie "this plant is backlit with early morning light, giving the flowers a cloud like appearance. In another location it is backlit at sunset, giving the flowers a beautiful golden hue. Proper site selection for this plant can really increase its dramatic value."

Hardiness Zone—Minimum: 3.

Planting Companions: Combines well with shorter perennials or shrubs. Larger grasses like *Calamagrostis*, *Miscanthus*, and *Panicum* are companions of comparable size, along with other large and bulky perennials such as *Echinops bannaticus* 'Blue Glow'.

Agastache 'Blue Fortune'
giant hyssop

Origin: Garden origin, a hybrid between a species from east Asia and one from North America.

Height: 40 inches.

Spread: 15 inches.

Description: "Sometimes," says Colleen Lockovitch, "you couldn't even see the flowers for butterflies." Perennial with masses of narrow, upright spikes of pale mauve blue gives this plant real presence. Particularly useful for contrasting with lower-growing, clump-forming species. The foliage has a distinct minty smell.

Season: Flowering July to September. Winter seed heads.

Garden Habitat and Cultivation: Any fertile, well-drained soil. Full sun. Can self-seed moderately.

Hardiness Zone—Minimum: 5.

Planting Companions: *Asclepias tuberosa*, *Echinacea* varieties, *Eryngium yuccifolium*, *Liatris spicata*, *Lythrum alatum*. "*Amsonia tabernaemontana* var. *salicifolia* makes for a wonderful combination too," notes Jennifer. "You just have to mind that the amsonia doesn't overgrow it."

Amorpha canescens

lead plant

Origin: Central states of Canada and the United States.

Height: 36 inches.

Spread: 30 inches.

Description: An oddity. Tiny, dull gray-purple flowers in tightly packed spikes and attractive gray-green divided leaves. An unusual plant, technically a dwarf shrub, with a rather awkward habit of growth.

Season: July to September.

Garden Habitat and Cultivation: Full sun, any reasonable soil, reasonably drought tolerant. The plant can be kept relatively tidy if it is treated as a perennial and cut back to ground level every winter. "Adored by rabbits," notes Colleen.

Hardiness Zone—Minimum: 2.

Planting Companions: *Echinacea* varieties, *Lythrum alatum*, *Persicaria amplexicaulis*, *Pycnanthemum muticum*, *Sanguisorba canadensis* 'Red Thunder'.

Amsonia

The blue stars are a distinctive group of perennials, most of them North American. The common name says it all for their main feature, but many

8.2. *Species of* Amsonia *are known as blue stars for their star-shaped flowers, often an unusually intense and steely blue in early to midsummer. This one is* Amsonia tabernaemontana *var.* salicifolia. *(Photo: Piet Oudolf)*

also have very good fall color, with a clear yellow rivaling that of some trees. All have upright stems with small leaves and form slowly expanding clumps. Although they are sometimes slow to establish, they are long-lived.

Planting Companions: *Asclepias tuberosa*, *Baptisia* spp., *Geranium* spp., *Liatris spicata*, *Parthenium integrifolium*, *Phlomis tuberosa* 'Amazone', *Sanguisorba menziesii*, *Stachys officinalis*.

Amsonia hubrichtii
Arkansas blue star

Origin: South central United States.

Height: 36 inches.

Spread: 24–36 inches, although spread may be slow.

Description: Dense heads of clear blue flowers atop very narrow foliage, which is a distinct feature of the plant. Colleen reports that it is a "fall icon plant, you can see the yellow leaves from condo buildings on Michigan Avenue."

Season: May to early June. Fall leaf color.

Garden Habitat and Cultivation: Found in a range of habitats in the wild, including full sun and dry shade. In cultivation, it does seem very drought tolerant. Less successful on heavier soils.

Hardiness Zone—Minimum: 5.

Amsonia tabernaemontana* var. *salicifolia
willowleaf blue star

Origin: Southeastern United States.

Height: 36 inches.

Spread: 36 inches, although spread may be slow.

Description: Heads of clear blue flowers with pale centers. Attractive pointed leaves turn yellow in fall. Dark stems are an attractive spring feature as the plants emerge from the ground. Older plants have an almost shrublike bulk.

Season: May to early June. Fall leaf color.

Garden Habitat and Cultivation: Tolerant of full sun and light shade. Has a tendency to self-seeding. Jennifer points out that "we remove seed pods every fall to reduce the amount of weeding, as they are hard to dig out once rooted."

Hardiness Zone—Minimum: 5.

Similar Varieties: *Amsonia tabernaemontana* 'Blue Ice' is very similar but with larger, darker flowers on a shorter plant. Jennifer enjoys how it "flowers while the foliage is developing and is a great combination for early season interest," adding with a note of relief that 'Blue Ice' "doesn't seem to seed like the others."

Anemone leveillei

no accepted common name

Origin: Central China.

Height: 24–30 inches.

Spread: 12 inches, slowly forming clumps.

Description: Good-sized white flowers with a pink- or blue-flushed exterior.

Season: June.

Garden Habitat and Cultivation: Best in sun or light shade with damp, well-drained soil. Easy and quick to establish.

Hardiness Zone—Minimum: 5.

Planting Companions: *Aruncus* 'Horatio', *Geranium* spp., *Hosta* spp., *Rodgersia pinnata* 'Superba', *Sanguisorba menziesii*, *Stachys officinalis*.

***Anemone x hybrida* and related species**

Japanese anemones

Origin: "Japanese" often means "introduced from Japanese nurseries," as Western plant hunters fell hungrily onto the country's highly developed nursery trade as soon as Commander Perry of the US Navy forced the country to open to foreign trade in 1853. This is a distinct group of woodland and woodland edge perennials from both China and Japan. The plants we have in cultivation are either varieties of *A. hupehensis*, *A. japonica*, or are hybrids of these and a third species, *A. vitifolia*.

Height: 60 inches.

Spread: 24 inches, continuing to spread to form large colonies.

Description: Large flowers shaped like shallow bowls facing outward. Robust-looking divided foliage. Very fluffy seed heads in fall. *Anemone*

8.3. ***The so-called Japanese anemones are very useful plants for bringing a long late season of color to lightly shaded areas. Slow to establish, they can live and slowly spread for many years. This one is*** **Anemone hupehensis.** ***(Photo: Noel Kingsbury)***

hupehensis 'Praecox' has bold pink flowers from early August onward; *Anemone hupehensis* 'Splendens', strong rosy-pink flowers. *Anemone x hybrida* 'Honorine Jobert' has large pure-white flowers—a superbly robust and beautiful plant, particularly useful for bringing light to shaded areas.

Season: "They start flowering August and September," says Colleen, "just when we need something, and they'll go on through November if we don't get a frost." Drought can also limit what is potentially a long flowering season.

Garden Habitat and Cultivation: Best grown in light shade, but without too much competition from tree roots, as these plants need moist, well-drained soils and reasonable fertility to do well. May suffer from sun scorch if planted in full sun in the south, but at northerly latitudes flourish in full sun. They can be slow to establish, but form very long-lived, spreading clumps with time; indeed, in some situations, they can become aggressive spreaders.

Hardiness Zone—Minimum: 4.

Planting Companions: *Aruncus* 'Horatio', *Astilbe chinensis var. taquetii* 'Purpurlanze'.

Eurybia divaricata and other *Symphyotrichum* (i.e., Aster) spp., *Eryngium yuccifolium*, *Eupatoriadelphus maculatus*, *Geranium x oxonianum* 'Claridge Druce', *Hosta* spp., *Pycnanthemum muticum*, *Tricyrtis* spp., *Veronicastrum virginicum*.

Aruncus 'Horatio'
goatsbeard

Origin: A hybrid bred by Ernst Pagels in northern Germany—one of the great plant breeders of the 1960s to 1990s. Ancestors from northern Asia and Europe.

Height: 50 inches.

Spread: 30 inches.

Description: Flowers, foliage—this one has it all. One of those plants that does not have an off day. Delicate cream flower clusters on bronze stems above divided golden-green leaves.

Season: Early June, plus good red color in fall and good structure in winter.

Garden Habitat and Cultivation: Easy in light shade, moist but well-drained soils. Slowly forms a clump, but not spreading. Does not appear to suffer from the leaf spotting, which can afflict some other goatsbeards.

Hardiness Zone—Minimum: 4.

Planting Companions: *Anemone leveillei*, *Anemone* Japanese types, *Astilbe chinensis var. taquetii* 'Purpurlanze', *Digitalis ferruginea*, *Eurybia divaricata*, *Geranium* spp., *Persicaria amplexicaulis*, *Polygonatum biflorum*, *Rodgersia pinnata* 'Superba'.

8.4. *The lively early summer cream flower clusters of a goatsbeard variety,* Aruncus *'Horatio', are matched by neat and elegant foliage. A robust and long-lived plant for light shade. This photograph was taken in the Oudolf garden. (Photo: Piet Oudolf)*

Asarum canadense
wild ginger
Origin: North America east of the Great Plains.
Height: 8 inches.
Spread: 20 inches, slowly spreading to form mats.
Description: Neat, rounded leaves to 4 inches across on spreading stems make this a useful ground-cover plant, although it may be slow to take off. The flowers are strange, brown, ground-level structures pollinated by flies—anyone so determined to find out what they smell like that they stick their nose to the ground is rewarded by an unpleasant odor.

8.5. *The intricately shaped midsummer flowers of swamp milkweed,* Asclepias incarnata, *are popular with a variety of insects. (Photo: Piet Oudolf)*

The roots were used by early settlers as a ginger substitute, hence the common name.

Garden Habitat and Cultivation: A plant of woodland, this needs light to full shade and a humus-rich, moist, but not waterlogged, soil.

Hardiness Zone—Minimum: 4

Planting Companions: *Dodecatheon meadia* 'Aphrodite', *Epimedium* spp., *Geranium x cantabrigiense* 'Karmina', *Helleborus x hybridus*, *Mertensia virginica*, *Polygonatum biflorum*.

Asclepias incarnata
swamp milkweed

Origin: Eastern United States.

Height: 60 inches.

Spread: 30 inches.

Description: Sometimes it pays to get out a magnifying glass—as with this milkweed; clusters of flowers varying between deep pink red and paler pinks reveal how intricately each individual flower is shaped. The clusters are atop upright stems with narrow leaves. Fluffy seed heads, but these are short-lived.

Season: July to August.

Garden Habitat and Cultivation: Full sun and average to wet soils. Self-seeding might be a problem in some gardens; on the other hand, the plant can be short-lived, so some seeding may be welcomed. Toxic white sap renders most milkweeds deer proof.

Hardiness Zone—Minimum: 3.

Planting Companions: *Astilbe chinensis* var. *taquetii* 'Purpurlanze', *Echinacea* spp., *Eryngium yuccifolium*, *Eupatoriadelphus maculatus*, *Parthenium integrifolium*, *Pycnanthemum muticum*, *Scutellaria incana*, *Veronicastrum virginicum*.

Asclepias tuberosa

butterfly weed

Origin: Eastern and southern United States.

Height: 24 inches.

Spread: 14 inches.

Description: Upright stems with clusters of yellow-orange flowers. The pointed leaves are food for monarch butterfly larvae. One of the most distinctive prairie species.

Season: June to August.

Garden Habitat and Cultivation: Full sun and medium to dry soils preferred, but not fussy about soil texture. It does need very good drainage, and plants do become diseased if overirrigated. Slow to establish.

Hardiness Zone—Minimum: 2.

Planting Companions: *Agastache* spp., *Baptisia* 'Purple Smoke', *Eryngium bourgatii*, *Eryngium yuccifolium*, *Gillenia trifoliata*, *Knautia macedonica*, *Nepeta* spp., *Salvia*—meadow sages, *Sedum* spp., *Vernonia lettermanii* 'Iron Butterfly'.

Aster divaricatus*—see *Eurybia divaricata

***Aster novae novae-angliae* 'Violetta'—see *Symphyotrichum novae-angliae* 'Violetta'**

***Aster tataricus* 'Jindai'**
Tatarican aster

Origin: Siberia to north China.
Height: 70 inches.
Spread: 30 inches.
Description: An upright aster with pale-violet daisy flowers with yellow centers. Unlike most American asters, the majority of the foliage is toward the lower part of the stems.
Season: October, so flowers may be frosted if grown farther north than Chicago.
Garden Habitat and Cultivation: Sun or light shade and average soils.
Hardiness Zone—Minimum: 3, possibly 4.
Planting Companions: *Anemone*—Japanese types, *Symphyotrichum oblongifolius* 'October Skies', *Persicaria amplexicaulis*, *Solidago rugosa* 'Fireworks'.

***Astilbe chinensis* var. *taquetii* 'Purpurlanze' and the very similar 'Purpurkerze' (purple candles), which is now being planted instead.**
purple lance astilbe

Origin: A selection of a species originally from eastern Asia.
Height: To 40 inches.
Spread: 30 inches, eventually forming large clumps.
Description: Masses of tiny and slightly fluffy lilac-pink flowers in upright clusters above divided leaves. Colleen says of it that "we keep it up all winter long, its shockingly good then as it holds its form so well."
Season: June to July. Winter seed heads.
Garden Habitat and Cultivation: Sun or light shade. While most astilbes need moist soils, this is the most tolerant of average conditions. However, it still resents drying out.
Hardiness Zone—Minimum: 4.
Planting Companions: *Anemone*—Japanese types, *Aruncus* 'Horatio', *Asclepias incarnata*, *Echinops bannaticus* 'Blue Glow', *Filipendula rubra* 'Venusta Magnifica', *Lythrum alatum*, *Parthenium integrifolium*, *Rodgersia pinnata* 'Superba', *Sanguisorba menziesii*.

8.6. **Astilbe chinensis** ***var.*** **taquetii** ***'Purpurlanze' has strikingly upright and uncompromisingly pink flower clusters, making a strong impression over a long period. Its winter seed heads are also striking. (Photo: Noel Kingsbury)***

Baptisia
wild indigoes

A group of members of the pea family, crucial to many prairie and savannah habitats, as they are immensely long-lived with huge root systems that reach to at least 10 feet underground. Like most pea family plants, they are able to "fix" atmospheric nitrogen, which makes them play an important role in prairie ecology—with other plants eventually benefiting from this vital nutrient. As with many plants with large root systems, they are slow to establish, but very long-lived once happy. They form tight clumps that spread only very slowly.

Baptisia 'Purple Smoke'
hybrid wild indigo

Origin: A natural hybrid introduced into cultivation by Niche Gardens, a nursery in North Carolina. Parents are eastern United States in distribution.

8.7. ***Wild indigo variety* Baptisia *'Purple Smoke' has flowers reminiscent of wisteria, followed by distinctive inflated seed pods. Its compact clusters of attractive foliage can be an important garden feature in their own right and are also valued by florists. (Photo: Robin Carlson)***

Height: 50 inches.
Spread: 40 inches.
Description: Smoky violet pea flowers in clusters above a neat, almost shrubby cluster of grayish leaves. Colleen thinks highly of it—"it's dynamite, another Lurie icon plant, everyone thinks it's asparagus when it comes up, the flowers are fine, but the silvery neat look of the leaves is there all summer long. . . . We called it the Lurie tumbleweed, because of the stems that break off and blow around in the winter."
Season: Late May, June. Winter seed heads.
Garden Habitat and Cultivation: Sun or light shade, any reasonable soil, including dry ones.
Hardiness Zone—Minimum: 4.
Planting Companions: *Echinacea* spp., *Geranium x oxonianum* 'Claridge Druce', *Liatris spicata*, *Monarda bradburyana*, *Phlomis tuberosa* 'Amazone', *Sanguisorba menziesii*. Grasses: *Sporobolus heterolepis*.

Baptisia alba var. _macrophylla_ (_B. leucantha_)
wild white indigo

Origin: Central and eastern North America.

Height: 50 inches.

Spread: 30 inches.

Description: Stiff, upright stems give this plant a dignified look, almost like a miniature tree; it can make a huge impact in the wild prairies where it grows. White flowers are spaced out on spikes above the foliage.

Season: June to July. Winter seed heads.

Garden Habitat and Cultivation: Full sun or light shade, seems tolerant of a range of soil moisture conditions.

Hardiness Zone—Minimum: 4.

Planting Companions: *Echinacea* spp., *Geranium x oxonianum* 'Claridge Druce', *Liatris spicata*, *Monarda bradburyana*, *Phlomis tuberosa* 'Amazone', *Sanguisorba menziesii*. Lower-growing grasses like *Sporobolus heterolepis*.

Calamintha nepeta subsp. _nepeta_
calamint

Origin: Southern and western Europe.

Height: 12 inches.

Spread: 20 inches.

Description: A tidily sprawling mass of wiry stems with small leaves and a characteristically minty smell (it is used in Italian cooking). The flowers are a pale lilac.

Season: July onward.

Garden Habitat and Cultivation: Very useful as a filler between taller plants, but without the risk of it spreading, as the central clump is actually very tight. Full sun, dry to medium soils. Jennifer is enthusiastic about what she says is "an amazing perennial—I love that the flowers are white when they begin blooming in the summer and then become a pale-violet color when the temperatures drop in early fall. I also love this plant because it feeds our bees for up to four months!"

Hardiness Zone—Minimum: 5.

8.8. ***Summer-flowering calamints (this is* Calamintha nepeta *subsp.* nepeta*) are very useful for filling spaces at the front and edges of borders or around the bases of larger and more upright plants. (Photo: Noel Kingsbury)***

Planting Companions: A favorite Piet Oudolf combination is with *Molinia* grasses, as the *Calamintha* will spread out between the tight grass clumps. This idea can be extended to using the plant between any other clumping relatively upright plants, with which its soft and vague outlines will be a good contrast.

Ceratostigma plumbaginoides
plumbago
Origin: Western China.
Height: 12 inches.
Spread: 15 inches.
Description: There are not many really true-blue flowers—this is one. A perennial with spreading stems, but in warmer climates the stems may persist and the plant adopt a low shrubby habit. Leaves turn red in fall.
Season: July to September. Fall color.

8.9. ***Purple prairie clover,* Dalea purpurea, *is a native plant of dry prairie habitats. Intensely colorful at midsummer, its growth is relatively slight, so it is important that it is not crowded by more vigorous plants. (Photo: Mark Tomaras)***

Garden Habitat and Cultivation: Full sun or light shade, any well-drained soil. The plants are often regarded as slow to establish.

Hardiness Zone—Minimum: 5.

Planting Companions: *Coreopsis verticillata* 'Golden Showers', *Echinacea* spp., *Heuchera villosa* 'Autumn Bride', *Origanum laevigatum* 'Herrenhausen', *Perovskia atriplicifolia* 'Little Spire'.

Dalea purpurea (Petalostemon purpureum)
purple prairie clover

Origin: Midwestern United States and Canada.

Height: To 24 inches.

Spread: 15 inches.

Description: Vivid magenta-purple flower heads make this one of the most intensely colored of midwestern wildflowers. Small, divided

8.10. ***Rusty foxglove,*** **Digitalis ferruginea,** ***is best grown in small clumps in order to appreciate its elegant upright habit. On cold winter mornings, its seed heads have a particularly mysterious and spectral quality. (Photo: Piet Oudolf)***

leaves scattered on upright stems. Like many other plants of dry habitats, it is slightly built and easily overwhelmed by more vigorous plants if conditions are fertile and moist.

Season: July to August.

Garden Habitat and Cultivation: Full sun. Medium to dry soils. Good on poor sandy soils. Very prone to rabbit damage.

Hardiness Zone—Minimum: 3.

Planting Companions: *Agastache* 'Blue Fortune', *Echinacea* spp., *Gentiana andrewsii*, *Geranium soboliferum*, *Liatris spicata*, *Limonium platyphyllum*, *Origanum laevigatum* 'Herrenhausen', *Ruellia humilis*, *Salvia*—meadow sages, *Stachys officinalis*, *Vernonia lettermanii* 'Iron Butterfly'.

Digitalis ferruginea
rusty foxglove

Origin: Southeast Europe.
Height: 60 inches.
Spread: 12 inches.
Description: Brown flowers! Yes, really! Although in fact, under close examination, they are an intricate mix of brown, fawn, and cream, only appearing brown at a distance. Numerous tubular flowers on a narrow spike does not make this plant sound very attractive, but the spike is so narrow the overall effect is of extraordinary elegance, especially if grown in groups. Each flower starts out peachy colored, darkening as it ages. The seed heads make a good winter feature too.
Season: Late June, July. Winter seed heads.
Garden Habitat and Cultivation: Sun or part shade, often thriving under trees. It is biennial or a short-lived perennial, but self-seeds well, even aggressively in some places.
Hardiness Zone—Minimum: 4.
Planting Companions: The tall and elegantly narrow flower and seed heads of this plant are highly effective grouped among any shorter species.

***Dodecatheon meadia* 'Aphrodite'**
shooting star

Origin: Nursery selection of a perennial native to eastern and central United States.
Height: 12 inches.
Spread: 8 inches.
Description: Unusual flowers with reflexed petals, a little like flower-shop cyclamen (to which it is very distantly related). Flower color is highly variable, from white to deep pink; this variety, however, has consistently large purple-pink blooms. Most effective in groups or scattered around in a border.
Season: April.
Garden Habitat and Cultivation: Sun or shade. May be naturalized in prairie plantings as it flowers very early, dying back to avoid the

8.11. *Shooting star variety* Dodecatheon meadia *'Aphrodite' is a garden form of a prairie native that pops up among the still-emerging shoots of much later developing plants in spring. It makes a cheerful and unusual addition to the garden but needs to be placed where it can be seen well enough to be appreciated. (Photo: Robin Carlson)*

competition of the taller and later-developing species. It is always summer dormant, so do not be surprised at a sudden disappearance!

Hardiness Zone—Minimum: 4.

Planting Companions: Colleen says that "they made the meadow a Disney film with blue *Eryngium yuccifolium* foliage and *Geum triflorum* seed heads. . . . The flowers look good with *Allium christophii* too."

Echinacea
coneflowers

The coneflower has become the poster boy for the native plant movement; *Echinacea purpurea*, in particular, with its big pink daisies, naturally has the show-off good looks breeders normally go to great lengths to give to their plants. However, there are problems! *E. purpurea* is always inclined to be short-lived and to suffer from disease. Needless to say, breeders have been hard at work making further improvements, particularly in widening the color range by crossing between various coneflower species.

In the wild, they can live for ten to twenty years, but their life span in the garden is often very much shorter than this. One reason is possibly that in gardens they are often crowded by other plants, and they appear not to like competition. Many of the modern varieties with flowers in the yellow and orange color range appear to be particularly short-lived.

Planting Companions: *Agastache* 'Blue Fortune', *Amsonia*, *Allium* 'Summer Beauty', *Amorpha canescens*, *Asclepias incarnata*, *Eurybia divaricata*, *Baptisia* spp., *Geranium* 'Brookside', *Lythrum alatum*, *Nepeta* spp., *Persicaria amplexicaulis*, *Ruellia humilis*, *Scutellaria incana*.

Echinacea pallida
pale purple coneflower

Origin: Eastern United States.

Height: 36 inches.

Spread: 18 inches.

Description: Pink daisy-type flower heads with narrow downward-pointing outer petals give this a dramatic appearance. In winter, the seed heads stand out as dark blobs against paler and wispier grasses. Like all the coneflowers, the foliage is a healthy-looking clump of

deep-green leaves. "It's a dream," says Colleen. "We leave it up all year; the finches loved the seed heads."
Season: June to July. Seed head interest in winter.
Garden Habitat and Cultivation: Sun to part shade. Any reasonable soil. Some drought tolerance. One of the more reliably long-lived and disease-free coneflowers.
Hardiness Zone—Minimum: 3.

Echinacea purpurea
purple coneflower
Origin: Eastern United States.
Height: 60 inches.
Spread: 20 inches.
Description: Robust, chunky flower heads with reflexed pink outer petals. The central cone of fertile florets is a rich orange-brown color.
Season: June to August. Some winter seed head interest.
Garden Habitat and Cultivation: Sun or light shade. Any reasonable soil. Some drought tolerance. May be short-lived and disease prone.
Hardiness Zone—Minimum: 3.
'Green Edge' has cream petals with a green edge. 'Rubinglow' is a rich carmine pink. 'Virgin', a variety introduced by Piet Oudolf, has white, horizontally held petals.

Echinacea tennesseensis
Tennessee coneflower
Origin: A rare species from Tennessee. Thought to be extinct for half a century until it was rediscovered in 1968 near Nashville. Unfortunately, this site was destroyed by the construction of a trailer park in the 1970s.
Height: 24 inches.
Spread: 18 inches.
Description: Purple-pink outer petals and green-coppery central cone. This species does not have the strongly reflexed petals of other wild coneflowers and has a particularly attractive shape, so what is quite a rarity naturally has ironically become important in cultivation as it has been used for breeding new hybrids.

Season: July to August.
Garden Habitat and Cultivation: Full sun and well-drained soil.
Hardiness Zone—Minimum: 5, but probably will take lower.

---

hybrid coneflowers

There are several other coneflowers grown in the Lurie, and no doubt more will be grown in the future. Most hybrids, however, do not appear to be as robust as the species or forms selected from the species, even though they make very colorful and rewarding garden plants. Many are quite weak at the base and appreciate the support of neighboring plants. As a very general rule, if there is yellow in the flower color (i.e., as orange or apricot), the more likely the variety is to be short-lived.

***Echinops bannaticus* 'Blue Glow'**
globe thistle

Origin: Nursery selection, species originally from southeast Europe.
Height: 50 inches.
Spread: 20 inches.
Description: Very distinctive, perfectly spherical steel-blue flower heads above coarse, thistle-like (but not prickly) foliage. The seed heads are unfortunately quite short-lived and soon shatter.
Season: July. Early fall seed head interest.
Garden Habitat and Cultivation: Full sun and any reasonable soil. Some gardeners find this a short-lived perennial; it may also self-seed. Prone to rabbit damage.
Hardiness Zone—Minimum: 3.
Planting Companions: *Aconongon* 'Johanneswolke', *Astilbe chinensis* var. *taquetii* 'Purpurlanze', *Eupatoriadelphus maculatus*, *Inula magnifica* 'Sonnenstrahl', *Lythrum alatum*, *Persicaria amplexicaulis*, *Pycnanthemum muticum*, *Veronicastrum virginicum*.

Epimedium
bishop's hat, barrenwort

Origin: Woodland plants from Asia and Europe. Species from Europe or North Africa are the most common in cultivation.
Height: 18 inches.

8.12. ***Intensely blue and perfectly spherical, the flower heads of globe thistle,* Echinops bannaticus *'Blue Glow'; midsummer flowering and popular with bees. (Photo: Noel Kingsbury)***

Spread: 18 inches, but will continue to spread indefinitely.

Description: Somewhat angular heart-shaped leaflets borne on stems that arise vertically from the ground forming a dense mat of foliage, thus making for decorative and robust ground cover. Small flowers of intriguing shape in clusters in spring. *Epimedium grandiflorum* 'Lilafee' (longspur barrenwort) has violet-pink flowers and purple-flushed young foliage in spring. *Epimedium x versicolor* 'Sulphureum' has coppery young foliage and pale-yellow flowers.

Season: Flowering in April to May, but the main feature is the foliage. In mild winter climates, some species are evergreen.

Garden Habitat and Cultivation: Shade; preferably humus-rich, moist but well-drained soil. Will tolerate dry conditions when established, but only in shade; may burn in full sun. Renowned as very long-lived plants, spreading slowly but surely.

Hardiness Zone—Minimum: 5.

Planting Companions: *Dodecatheon meadia* 'Aphrodite', shorter *Geranium*

8.13. *Species of barrenwort (*Epimedium*) look particularly good as their leaves emerge in spring but are valuable as summer-long ground cover for lightly shaded places. In areas with milder winters, many species are evergreen. Here in the Dark Plate is* Epimedium grandiflorum *'Lilafee', whose purple-flushed leaves and curiously shaped lilac flowers show off daffodil* Narcissus *'Lemon Drops'. The blue is* Mertensia virginica. *(Photo: Robin Carlson)*

spp., *Geum rivale* 'Flames of Passion', *Jeffersonia diphylla, Mertensia virginica, Polygonatum biflorum.*

Eryngium bourgatii

eryngo

Origin: Spain, Portugal, northwest Africa.

Height: 22 inches.

Spread: 15 inches.

Description: A perennial with a distinct desert look. Very distinctive gray-green leaves, with white markings. The flower head is thimble-shaped and surrounded with painfully sharp spines. Good forms have a rich blue coloring but can look untidy after flowering.

Season: July. Foliage decorative May to October.

Garden Habitat and Cultivation: Full sun. Tolerant of drought and poor soils. Dislikes wet soils. Cutting back after flowering encourages production of smart young foliage.

Hardiness Zone—Minimum: 5.

Planting Companions: *Asclepias tuberosa, Dodecatheon meadia* 'Aphrodite', *Gentiana andrewsii, Geranium sanguineum* 'Max Frei', *Knautia macedonica, Limonium platyphyllum, Nepeta* spp., *Origanum laevigatum* 'Herrenhausen', *Salvia*—meadow sages, *Vernonia lettermanii* 'Iron Butterfly'.

Eryngium 'Big Blue' is a form, or possibly a hybrid, of *E. bourgatii*, with iridescent blue flowers on upright stems. It is slightly taller, at 30 inches.

Eryngium yuccifolium
rattlesnake master

Origin: Midwestern and southeastern United States.

Height: 60 inches.

Spread: 24 inches.

Description: Gray-white spherical flower heads atop spiny-looking foliage, slightly reminiscent of a yucca. Not hugely colorful, but it has good structure.

Season: June to September. Some winter structural interest.

Garden Habitat and Cultivation: Full sun or light shade, medium to dry soils. Can self-seed strongly, and so perhaps best in a meadow or prairie setting where grasses and other plants can keep it in check.

Hardiness Zone—Minimum: 3.

Planting Companions: *Agastache* 'Blue Fortune' and *A. rupestris*, *Anemone*—Japanese types, *Asclepias incarnata, and A. tuberosa*, *Eupatoriadelphus maculatus*, *Perovskia atriplicifolia* 'Little Spire', *Persicaria amplexicaulis*, *Veronicastrum virginicum.*

Eupatoriadelphus maculatus 'Gateway'
(formerly _Eupatorium maculatum_ 'Gateway')
joe-pye weed

Origin: Nursery selection of a species to be found across southern Canada and midwestern, northeastern, and mountain states of the United States. There are several other very similar species: *Eupatoriadelphus fistulosus* and *Eupatorium pupureum*—the naming is very confusing!

Height: To 60 inches.

Spread: 20 inches.

Description: A majestic perennial with upright stems and large, flattish dull-pink flower heads, adored by butterflies. This variety is shorter than the wild species. 'Purple Bush' is very similar but shorter and often better in smaller gardens.

Season: July to September.

8.14. *The joe-pye weeds are familiar from many wild habitats but also make very good garden plants. This is* Eupatoriadelphus maculatus *'Atropurpureum' in the garden of designer Piet Oudolf flowering in midsummer, very similar to the variety 'Gateway' in the Lurie Garden. In front of it are* Persicaria amplexicaulis *'Rosea' (*left*) and a white form of* Veronicastrum virginicum *(*right*).This planting is over twenty years old, and apart from weeding and annual cutting back, has required minimal maintenance and no replanting. (Photo: Noel Kingsbury)*

Garden Habitat and Cultivation: Full sun, fertile and moist soil needed for success. Wilts badly in dry conditions.
Hardiness Zone—Minimum: 4.
Planting Companions: *Anemone*—Japanese types, *Asclepias incarnata*, *Symphyotrichum novae-angliae*, *Echinops bannaticus* 'Blue Glow', *Eryngium yuccifolium*, *Persicaria amplexicaulis*.

Eurybia divaricata (_Aster divaricatus_)
white wood aster
Origin: Southeastern Canada, eastern United States.
Height: 24 inches.
Spread:12 inches, but can form large clumps with time.

Description: White asters never seem as ornamental as the blue ones, but the masses of white flowers in shade where little else flowers in late summer are a much-appreciated sight. A close examination reveals elegant, dark, wiry stems supporting the flowers.
Season: August to fall.
Garden Habitat and Cultivation: Sun or shade, with a tolerance of dry soils. Self-sowing and aggressive spread are possible in some gardens.
Hardiness Zone—Minimum: 3.
Planting Companions: *Anemone*—Japanese types, *Aruncus* 'Horatio', *Echinacea* spp., *Eupatoriadelphus maculatus*, *Gentiana andrewsii*, *Geranium* 'Jolly Bee', *Heuchera villosa* 'Autumn Bride', *Persicaria amplexicaulis*.

Filipendula rubra 'Venusta Magnifica'
queen of the prairie
Origin: Eastern half of the United States.
Height: 100 inches.
Spread: 40 inches.
Description: With a cloud of pale-pink, fluffy flowers held above divided foliage, this is a striking and distinctive plant.
Season: All too briefly in June.
Garden Habitat and Cultivation: Full sun and fertility important, but moist soil is absolutely essential. Very prone to flopping or browning during drought. Jennifer says that "we just cut it back half way or cut half the stems to the ground to avoid it flopping. It then produces new growth and looks fabulous."
Hardiness Zone—Minimum: 3.
Planting Companions: *Astilbe chinensis* var. *taquetii* 'Purpurlanze', *Baptisia* spp. *Phlomis tuberosa* 'Amazone', *Veronicastrum virginicum*.

Gentiana andrewsii
bottle gentian
Origin: Eastern North America.
Height: 24 inches.
Spread: 12 inches.
Description: Vivid blue flowers sit in tight clusters atop erect stems; that

8.15. *The flowers of* Gentiana andrewsii *seem so full of promise—they always look like they are about to open, but never do. A slow-growing but long-lived prairie species. (Photo: Robin Carlson)*

the flowers never actually open properly might seem a little strange, but it does not detract from their beauty.

Season: Flowers in August to September, followed by yellow fall color.

Garden Habitat and Cultivation: Light shade is best, in a humus-rich soil, moist but well drained. Slow to establish but long-lived; "just a winner," says Colleen. Jennifer thinks highly of it too—"this is one of the fall highlights in the garden, the stunning blue flowers are spectacular on their own, but watching the bees open up the petals to collect pollen and then reverse out of the tube-shaped flowers always stops visitors in their tracks."

Hardiness Zone—Minimum: 3.

Planting Companions: *Eurybia divaricata*, *Dalea purpurea*, *Eryngium bourgatii*, *Heuchera villosa* 'Autumn Bride', *Sedum* spp., *Tricyrtis* spp. Jennifer suggests "*Calamintha nepeta* subsp. *nepeta*, as this is the perfect combo as the calamintha holds up the gentian blooms and it looks fantastic."

Geranium 'Brookside'
a variety of cranesbill

Origin: Nursery origin, parents from Europe and Asia.

Height: 18 inches.

Spread: 18 inches, slowly forming a spreading clump.

Description: Clear violet-blue flowers with a pale eye smother a mounding plant. One of the best blue cranesbills. Foliage can sometimes turn a nice orange in fall. The variety 'Orion' is similar but larger.

Season: June, and sometimes repeat flowering for the rest of the summer.

Garden Habitat and Cultivation: Sun or light shade, any reasonable soil. Some drought tolerance if out of direct sunlight.

Hardiness Zone—Minimum: 4.

Planting Companions: *Echinacea* spp., *Geranium sanguineum* 'Max Frei', *Nepeta* spp., *Sanguisorba menziesii.*

Geranium x cantabrigiense 'Karmina'
a variety of cranesbill

Origin: Garden hybrid between two European species.

Height: 8 inches.

Spread: 12 inches.

Description: Small lilac-pink flowers held above slightly glossy foliage, which in combination with running stems at ground level ensures a low, neat ground-cover effect.

Season: May to June.

Garden Habitat and Cultivation: Light shade to full sun, any reasonable soil. Some tolerance of dry shade. It can react badly to overgenerous

8.16. ***Cranesbill (*Geranium x cantabrigiense *'Karmina') shows off its potential as a ground-cover plant in the Dark Plate. (Photo: Robin Carlson)***

irrigation. Its steadily running stems give the capacity for continued spread, but this is never a problem.

Hardiness Zone—Minimum: 5.

Planting Companions: *Geranium sanguineum* 'Max Frei', *Nepeta* spp., *Sanguisorba menziesii*. Most effectively used planted in multiples as a ground cover.

Geranium 'Dilys'
a variety of cranesbill

Origin: Nursery origin, parents from Europe and Asia.

Height: 12 inches.

Spread: 30 inches, forming a slowly spreading clump.

Description: Pale-purple flowers with darker veining.

Season: June, and sometimes repeat flowering later in the summer, especially if cut back after the initial flowering.

8.17. **Geranium *'Jolly Bee' is a variety of cranesbill, which flowers for months from midsummer and makes very effective summer ground cover. (Photo: Noel Kingsbury)***

Garden Habitat and Cultivation: Sun or light shade, any reasonable soil.
Hardiness Zone—Minimum: 4.
Planting Companions: *Echinacea* spp., other *Geranium*, *Nepeta* spp., *Sanguisorba menziesii*.

Geranium 'Jolly Bee'
a variety of cranesbill
Origin: Garden hybrid between two Asian species.
Height: 24 inches.
Spread: To 30 inches.
Description: Rounded petals of pale blue with a dash of violet and central pale zone smother a sprawling (but tidy) mass of stems and leaves.
Season: July to September—and exceptionally reliable in doing this, thus making it beloved of garden center managers.
Garden Habitat and Cultivation: Part shade to full sun, any reasonable

soil. The shape of the plant lends itself to well-behaved ground cover, in that, although it spreads out over a wide distance, this is because the stems all radiate out from one central growth point, not because the plant itself is spreading—in contrast to most geraniums, which are spreading.

Hardiness Zone—Minimum: 4.

Planting Companions: *Eurybia divaricata*, *Coreopsis verticillata* 'Golden Showers', *Hemerocallis* varieties, *Heuchera villosa* 'Autumn Bride', *Persicaria amplexicaulis*, *Sedum* spp.

Geranium x oxonianum 'Claridge Druce'
a variety of cranesbill

Origin: A garden hybrid between two southern European species. Claridge Druce was an early twentieth-century pharmacist in the university town of Oxford famous for hangover cures.

Height: 30 inches.

Spread: 24 inches.

Description: Strong pink flowers atop a mound of slightly coarse-looking foliage.

Season: June, but can repeat flower in September.

Garden Habitat and Cultivation: Light shade to full sun, any reasonable soil. After flowering, the plant can look very scruffy—cutting back now will encourage the production of a second set of flowers for late summer. Potentially an extremely vigorous plant, its clumps spreading at a moderate pace, and often self-seeding. However, it is not reliably frost-hardy, but where it is, very useful as a low-maintenance plant for lightly shaded areas.

Hardiness Zone—Minimum: 5.

Planting Companions: *Anemone*—Japanese types, *Baptisia* spp., *Geranium* 'Brookside', *Gillenia trifoliata*, *Nepeta* spp., *Persicaria amplexicaulis*, *Rodgersia pinnata* 'Superba', *Stachys officinalis*, *Veronicastrum virginicum*.

Geranium sanguineum 'Max Frei'
a variety of bloody cranesbill

Origin: Europe.

Height: 8 inches.

8.18. ***A group of*** **Geranium x oxonianum** ***varieties. 'Claridge Druce' is one of the larger and more vigorous of this extensive group of cranesbills. Varying in color from almost white to vivid magenta, they are not always hardy in the Midwest but are worth persisting with, as they often repeat flower, recover rapidly from drought, and are tolerant of shade. The numbers of*** **G. x oxonianum** ***in cultivation continue to increase—some of these will be more tolerant of cold than others. (Photo: Noel Kingsbury)***

Spread: 18 inches, spreading steadily to form large clumps.

Description: Large and striking magenta flowers over small and attractive divided leaves. Not for those who like subtlety in the garden.

Season: May to June, some flowering in later months.

Garden Habitat and Cultivation: Full sun or light shade, any reasonable soil including poor and dry ones.

Hardiness Zone—Minimum: 4.

Planting Companions: *Eryngium bourgatii*, *Geranium* 'Brookside', *Geranium x cantabrigiense* 'Karmina', *Lythrum alatum*, *Nepeta* spp., *Salvia*—meadow sages, *Stachys officinalis*, *Vernonia lettermanii* 'Iron Butterfly'.

Geranium soboliferum

a cranesbill species

Origin: Far eastern Russia, Japan, northern China.

Height: 24 inches.

Spread: 24 inches, spreading steadily to form large clumps.

Description: Moderately magenta flowers over deeply cut leaves that turn crimson in fall, making it one of the most striking fall perennials.

Season: September to October.

Garden Habitat and Cultivation: Full sun or light shade, any reasonable well-drained soil.

Hardiness Zone—Minimum: 5.

Planting Companions: *Dalea purpurea*, *Gentiana andrewsii*, *Origanum laevigatum* 'Herrenhausen', *Symphyotrichum* 'Blue Autumn' and *S. oblongifolius* 'October Skies',.

***Geum rivale* 'Flames of Passion'**

wood avens

Origin: A nursery selection from a species found across Canada, northern United States, and northern Europe.

Height: 15 inches.

Spread: 15 inches, slowly forming extensive clumps.

Description: Deep dark-red double flowers above neat clumps of broad foliage. A Piet Oudolf selection.

Season: May.

8.19. **Geranium soboliferum** *is later flowering than many cranesbill geraniums and so makes a useful lower-level plant for August and September interest. Its brightly colored fall foliage looks its best just before the first hard frosts. Several other east Asian* **Geranium** *species put on a good fall show too. (Photo: Piet Oudolf)*

Garden Habitat and Cultivation: Full sun or light shade, any reasonable soil.
Hardiness Zone—Minimum: 3.
Planting Companions: *Mertensia virginica*, *Nepeta* spp., *Salvia*—meadow sages, *Zizia aurea*.

Geum triflorum
prairie smoke, old man's whiskers
Origin: Great Plains westward, plus the most northerly states and Canada.
Height: 16 inches.

8.20. **Geum triflorum** ***may be shyly undistinguished in flower but makes up for it with a spectacular early to midsummer display of fluffy seed heads. (Photo: Robin Carlson)***

Spread: 12 inches.

Description: Small, murky pink-red flowers in May are followed by fluffy seed heads in July. There is nothing quite like one of the dry prairies where this plant grows for the spectacle of thousands of these seed heads, an effect that is easily reproduced on a small scale in the garden.

Season: July.

Garden Habitat and Cultivation: Full sun, drier soils. In the Lurie Garden, it is used very successfully along border edges. Surprisingly, it does self-seed particularly strongly. It is important that it is not crowded by other plants, especially taller species.

Hardiness Zone—Minimum: 3.

Planting Companions: *Dodecatheon meadia* 'Aphrodite', *Eryngium bourgatii*, *Nepeta* spp., *Salvia*—meadow sages, *Vernonia lettermanii* 'Iron Butterfly'.

Gillenia trifoliata (Porteranthus trifoliata)

bowman's root

Origin: Eastern United States and southeast Canada.

Height: 30 inches.

Spread: 15 inches.

Description: One of those plants that does not sound very exciting but always captivates people. Small white flowers on wiry red stems above neat fresh green leaves. An essay in elegance. Colleen describes it as "dynamite in the fall, the color (red) is out of this world." A good cut flower, lasting weeks in a vase.

Season: Late May to June. Fall color.

Garden Habitat and Cultivation: Light shade to full sun. Moist but well-drained soils, preferably with a high humus content. Dislikes drought.

Hardiness Zone—Minimum: 4.

Planting Companions: *Asclepias tuberosa*, *Geranium x oxonianum* 'Claridge Druce', *Lythrum alatum*, *Sanguisorba menziesii*, *Stachys officinalis*, *Solidago rugosa* 'Fireworks'.

Helleborus x hybridus (often incorrectly _H. orientalis_)

lenten rose

Origin: Hybrids derived from species originally from southeast Europe and Turkey.

Height: 16 inches.

Spread: 20 inches.

Description: If this flowered in summer, we probably would not bother growing it, but in early spring we are all bewitched by wide-open, bowl-shaped flowers in an extraordinary range of colors: creams, pale yellows, purples, purple blacks, pinks, sometimes with markings. The dark-green, leathery foliage is divided into leaflets radiating out from a central point—it is very attractive in its own right, especially since it lasts for a whole year and is therefore effectively evergreen.

Season: March to April.

Garden Habitat and Cultivation: Colleen notes that "it does not really need shade, seems OK in full sun," despite it being a woodland plant in its native country at a latitude shared by much of the continental United States. Plants thrive in any well-drained reasonable soil and do well enough on poor ones, but they do better without summer drought. The foliage overwinters well, but at flowering time is generally replaced by fresh leaves. Plants may self-seed strongly—seedlings can be transplanted when young but cannot be guaranteed to have the

8.21. *While* Helleborus *hybrids have long been in cultivation, recent years have seen a great increase in the number of varieties, sometimes with unusual patterns of spotting or in dark and mysterious colors. Flowering soon after the end of winter, it is tempting to cut them for the house, but they wilt almost immediately; cut and floated in water, however, they last for up to a week. (Photo: Noel Kingsbury)*

same flower characteristics of the parent. Seed is really the only practical way to propagate it, as clumps form only very slowly; it should be sown as soon as it is ripe, although it may not germinate until the next spring.

Hardiness Zone—Minimum: 4.

Planting Companions: With their distinctive broad foliage, hellebores can be shoehorned in between a wide variety of other perennials.

Hemerocallis
day lilies

These are among the best loved of all perennials and the most truly American, although in fact their ancestors are from Eurasia. They flourish in hot humid summers, producing a succession of trumpet-shaped flowers, each one of which only lasts for one day. Almost absurdly easy to breed (every seedling from a cross looks good!), there are around sixty thousand cultivars recognized by the American Hemerocallis Society—they long ago ran out of sensible names for new plants.

All flourish in full sun or light shade and any reasonable soil; some drought and occasional waterlogging rarely deter them. Vigor, size (especially height), and winter hardiness may vary between varieties. Only two have been used in the Lurie Garden (at the time of writing): 'Chicago Apache' and 'Gentle Shepherd', both of which have performed extremely well—information given here is based on these two varieties.

Height: 30 inches.
Spread: 24 inches.
Description: 'Gentle Shepherd' has flared petals of palest yellow. 'Chicago Apache' is a stunning rich scarlet.
Season: July to August.
Hardiness Zone—Minimum: 3. Some varieties are notably less hardy.
Planting Companions: *Geranium* 'Jolly Bee', *Knautia macedonica*, *Parthenium integrifolium*, *Pycnanthemum muticum*, *Rodgersia pinnata* 'Superba', *Ruellia humilis*, *Scutellaria incana*.

Heuchera villosa 'Autumn Bride'
hairy alum root, maple leaf alum root, coral bells

Origin: A nursery selection of a species native to the eastern United States.
Height: 30 inches.
Spread: 20 inches.
Description: The common name—maple leaf—says it all: big, fresh green leaves held horizontally on a clump-forming plant. The flowers are tiny and white and massed into upright spikes.
Season: August to September.
Garden Habitat and Cultivation: Light to full shade; humus-rich, moist

8.22. *The mapleleaf alumroot (*Heuchera villosa *'Autumn Bride') thrives in light shade and, like most heucheras, is grown for its fine-looking foliage rather than flowers. Varieties with brightly colored leaves are also increasingly popular. (Photo: Robin Carlson)*

but well-drained soil. The plants form slowly spreading clumps with shoots and roots growing out of thick, almost woody horizontal stems—with time, older parts die off; plants may need occasional division and replanting to ensure the survival of vigorous younger parts.

Hardiness Zone—Minimum: 3.

Planting Companions: *Ceratostigma plumbaginoides, Eurybia divaricata, Geranium* 'Jolly Bee'.

Heuchera richardsonii is similar but about two-thirds the size. There are many other varieties, mostly bred for brightly colored foliage.

Hosta
hostas

Hostas are the premier garden foliage perennials, available in a vast number of forms, differing primarily in leaf color and patterning, but also in height and vigor. Originally from the Far East, it was Chinese and Japanese gardeners who made the first garden selections of these broad-leaved plants over two centuries ago. The flowers are also attractive, and often beautifully scented, being held on spikes considerably above leaf height.

Cultivation needs are generally similar—humus-rich moist to wet (but never stagnant) soil in light to full shade is essential. Most varieties also need protection from slug and snail damage. The plants form long-lived clumps that slowly expand—with time they can make impressive and weed-proof ground cover.

In the Lurie Garden the following varieties are used: 'Blue Angel', 'Halcyon', 'Krossa Regal', 'Royal Standard', and 'White Triumphator'. Details are given for these as a group, and are broadly typical of the many varieties in common cultivation.

Height: 24 to 40 inches (at flowering).

Spread: 30 to 40 inches.

Description: 'Blue Angel', 'Halcyon', and 'Krossa Regal' have gray-blue leaves with distinctive veining; 'Royal Standard' and 'White Triumphator' have lush green foliage. Flower color varies from white to pale lilac; 'White Triumphator' has scented white flowers on tall stems (40 inches) that are among the best.

Season: The foliage, from June onward, is the main interest. Flowering times vary (itself a valuable characteristic) from June to August.

Hardiness Zone—Minimum: 3.

Planting Companions: As distinctive foliage plants, hostas fit in well with a wide variety of other perennials—which is a big part of their appeal.

Inula magnifica 'Sonnenstrahl'
a species of elcampagne

Origin: A German nursery selection made of a species from the Caucasus region, between Russia and Iran.

Height: 80 inches.

Spread: 30 inches.

Description: One of those plants that tends to make those who have never seen it before go "wow"—an impressive accent plant at all stages of growth. Very large, coarsely hairy leaves with stout erect stems producing heads of golden-yellow daisy flowers, notable for their very fine ray florets.

Season: August.

Garden Habitat and Cultivation: Full sun and fertile soil. The plants appreciate damp conditions, but this particular variety is noted as the most drought tolerant.

8.23. *The big bright yellow daisies of* Inula magnifica *'Sonnenstrahl' top tall stems in midsummer. Wind can damage them easily, so shelter or staking is advisable. (Photo: Piet Oudolf)*

Hardiness Zone—Minimum: 5.
Planting Companions: *Echinops bannaticus* 'Blue Glow', *Persicaria amplexicaulis, Sanguisorba canadensis* 'Red Thunder'.

Jeffersonia diphylla
twinleaf
Origin: Eastern North America.
Height: 18 inches.
Spread: 10 inches.
Description: Delicate pure-white flowers that seem to sum up the ephemeral feeling of spring. The leaves are very unusual, divided into two very distinct halves. Although the flowers are short-lived, this can be a valuable summer foliage plant for shade.
Season: April.
Garden Habitat and Cultivation: Shade with humus-rich soil, moist but well drained. From alkaline soils in nature, but happy with moderately

8.24. *The color of scabious (*Knautia macedonica*) flowers is quite unlike anything else and often astonishes those who see it for the first time. There is variation, however, from a dark red to a pale cerise pink. (Photo: Noel Kingsbury)*

acidic. Growth in the first few years will be slow, but then it can be expected to gently run around and between other plants.

Hardiness Zone—Minimum: 4.

Planting Companions: *Dodecatheon meadia* 'Aphrodite', *Epimedium* spp., *Helleborus x hybridus, Mertensia virginica.*

Knautia macedonica

knautia

Origin: Southeast Europe.

Height: 30 inches.

Spread: 20 inches.

Description: Button-shaped flower heads of extraordinary intensity, neither pink nor red but somehow both, on long, rangy stems. The foliage tends to form tight clumps near ground level.

8.25. *The Midwest prairie native blazing star,* Liatris spicata, *is popular with the cut-flower industry for its long-lasting and colorful mid- to late-summer flowers. (Photo: Noel Kingsbury)*

Season: July, but with reasonable soil moisture, this plant can carry on flowering until September

Garden Habitat and Cultivation: Full sun, any well-drained soil, including poor ones. It is always short-lived—five years at the most, but often self-seeds prolifically. If cut back after the first flowering, it repeat flowers.

Hardiness Zone—Minimum: 4.

Planting Companions: *Asclepias tuberosa*, *Eryngium bourgatii*, *Hemerocallis* varieties, *Nepeta* spp., *Perovskia atriplicifolia* 'Little Spire', *Ruellia humilis*.

Liatris spicata

blazing star

Origin: Eastern United States.

Height: 45 inches.

Spread: 12 inches.

Description: Spikes of very distinctive, fluffy-looking purple-pink flowers that start to open from the top down. Linear foliage. Despite bearing no obvious resemblance, a member of the daisy family. Highly regarded as a cut flower. There are several varieties available commercially, including 'Alba', which is white.

Season: June, July.

Garden Habitat and Cultivation: Full sun; any reasonable soil, including poor ones. Although relatively drought tolerant, it performs better with reasonable levels of soil moisture during the growing season, but too much winter damp can cause rotting of the bulblike roots. Prone to rabbit damage.

Hardiness Zone—Minimum: 3.

Planting Companions: *Agastache* 'Blue Fortune', *Amsonia* spp., *Baptisia* spp., *Dalea purpurea*, *Lythrum alatum*, *Nepeta* spp., *Stachys officinalis*, *Vernonia lettermanii* 'Iron Butterfly'.

Limonium platyphyllum (L. latifolium)

sea lavender

Origin: Southeast Europe.

Height: 30 inches.

Spread: 20 inches.

8.26. ***Clouds of sea lavender,*** **Limonium platyphyllum.** ***This very drought-tolerant species has a long season of cloudlike flower heads, which continue to look attractive long after the tiny flowers have died. (Photo: Piet Oudolf)***

Description: Large glossy paddle-shaped leaves form a rosette at ground level with sprays of tiny lilac flowers in cloudlike heads held well above the foliage.

Season: September to October.

Garden Habitat and Cultivation: Full sun, any well-drained soil. Notably drought tolerant. Long-lived and reliable.

Hardiness Zone—Minimum: 3.

Planting Companions: *Coreopsis verticillata* 'Golden Showers', *Dalea purpurea, Eryngium bourgatii, Origanum laevigatum* 'Herrenhausen', *Perovskia atriplicifolia* 'Little Spire', *Sedum* spp., *Vernonia lettermanii* 'Iron Butterfly'.

Lythrum alatum
loosestrife

Origin: Central and eastern United States, eastern Canada.

Height: 36 inches.

Spread: 24 inches.
Description: Upright growing with pink flowers.
Season: May to September.
Garden Habitat and Cultivation: Full sun, any reasonable soil but most successful on moister soils. May self-sow. It is closely related to one of the most notorious invasive aliens—purple loosestrife (*Lythrum salicaria*), a European species, now illegal to plant in some states. *L. alatum* is not invasive and is a much more delicate-looking plant.
Hardiness Zone—Minimum: 3.
Planting Companions: *Agastache* 'Blue Fortune', *Amorpha canescens, Astilbe chinensis* var. *taquetii* 'Purpurlanze', *Echinacea* spp., *Echinops bannaticus* 'Blue Glow', *Liatris spicata, Rodgersia pinnata* 'Superba', *Scutellaria incana.*

Maianthemum racemosa (Smilacina racemosa)
false Solomon's seal, false spikenard
Origin: Temperate North America.
Height: 36 inches.
Spread: 18 inches.
Description: White plumes of flowers above rather fine dark-green leaves on erect stems.
Season: April, May, but flowering season short. Red berries and yellow foliage in fall.
Garden Habitat and Cultivation: A woodlander, so must have shade or light shade, and humus-rich, moist but well-drained soil. If it likes you, it will form large clumps with time.
Hardiness Zone—Minimum: 3.
Planting Companions: A short flowering season limits its impact, but its leaves look good alongside other woodland species.

Mertensia virginica
Virginia bluebells
Origin: Eastern United States and Canada.
Height: 24 inches.
Spread: 18 inches.
Description: Ethereally pale-blue flowers of a tubular bell shape, hanging down from lush-looking leaves. A delightful spring woodland flower.

8.27. ***Virginia bluebell,*** **Mertensia virginica,** ***is a native woodland wildflower that can be left to grow in among other plants, flowering in spring but, like a bulb, dying back in the heat of the summer. (Photo: Robin Carlson)***

Season: May.

Garden Habitat and Cultivation: Shade and humus-rich, moist but well-drained soil. Unlike many woodlanders, this is easy and quick to establish, the underground roots slowly spreading. The foliage is summer dormant and care must be taken not to disturb the plant during this time.

Hardiness Zone—Minimum: 3.

Planting Companions: *Dodecatheon meadia* 'Aphrodite', *Epimedium* spp., *Geranium x cantabrigiense* 'Karmina', *Helleborus x hybridus, Mertensia virginica, Polygonatum biflorum, Zizia aurea.* Jennifer thinks that "Virginia bluebells blooming alongside narcissus and the red new foliage of

8.28. ***Attractive to flower arrangers as much as bees, the midsummer-flowering eastern bee balm (*Monarda bradburyana*) has a pleasant herbal aroma and was used by early settlers for tea. (Photo: Piet Oudolf)***

Penstemon 'Husker's Red' is a stunning combination in spring—one of my favorites."

Monarda bradburyana
eastern bee balm, white bergamot
Origin: A band reaching from Texas to the Great Lakes.
Height: 18 inches.
Spread: 12 inches.
Description: Very pale lavender-pink spotted flowers in whorled clusters with distinct brown bracts. Unusual and looks good in a vase. Jennifer reckons it to be "one of the best plants for conveying the concept of four season interest—nice green foliage in the spring, fantastic pink blooms in early summer, gorgeous seed heads through fall, and fall

foliage color in a deep maroon." There are many other varieties of *Monarda* commercially available that make colorful midsummer flowers for the border; often, however, they are difficult to keep going from year to year without digging up and replanting.

Season: May to June.

Garden Habitat and Cultivation: Full sun, light shade, any reasonable soil, some drought tolerance. It forms loose clumps. Mildew resistant compared to other varieties of *Monarda*.

Hardiness Zone—Minimum: 4.

Planting Companions: *Amsonia* spp., *Baptisia* spp., *Echinacea* spp., *Liatris spicata*, *Phlomis tuberosa* 'Amazone', *Sanguisorba menziesii*, *Stachys officinalis*. Particularly effective with grass *Sporobolus heterolepis*.

Nepeta racemosa 'Walker's Low'

a species of catmint

Origin: A garden selection of a European species.

Height: 30 inches.

Spread: 30 inches, forming a dense slowly expanding clump.

Description: Profuse violet-blue flowers on a sprawling plant with densely packed, small gray leaves. The whole plant is very aromatic; the plant appears to be an intoxicant for cats—but only some cats get addicted! The plant's habit makes it very useful for ground cover or bank planting, although the well-known N. x *faassenii* is actually better for this.

Season: May to June; if dead spikes are cut back and the plant is kept moist, repeat flowering is possible until September.

Garden Habitat and Cultivation: Full sun or light shade, any reasonable soil, plus poor ones, some drought tolerance.

Hardiness Zone—Minimum: 4.

Planting Companions: *Asclepias tuberosa*, *Echinacea* spp., *Eryngium bourgatii*, *Geranium* 'Brookside', *Geranium x cantabrigiense* 'Karmina', *G. x oxonianum* 'Claridge Druce', *G. sanguineum* 'Max Frei', *Geum rivale* 'Flames of Passion', *G. triflorum*, *Knautia macedonica*, *Liatris spicata*, *Sanguisorba menziesii*, *Stachys officinalis*.

8.29. **Nepeta racemosa *'Walker's Low', a variety of catmint, is useful for the edges of borders or growing on slopes, as its stems gently sprawl over the ground. (Photo: Piet Oudolf)***

Nepeta subsessilis 'Sweet Dreams'
a variety of catmint

Origin: A nursery selection of a plant from Japan.

Height: 24 inches.

Spread: 18 inches.

Description: Plentiful violet-blue, elongated flowers on spikes above densely packed, small leaves.

Season: May to June; if dead spikes are cut back and the plant is kept moist, repeat flowering is possible until September. Good fall color.

Garden Habitat and Cultivation: Sun or shade, any reasonable soil that does not dry out. Can be slow to establish.

Hardiness Zone—Minimum: 4.

Planting Companions: See above.

***Origanum laevigatum* 'Herrenhausen'**

a species of oregano

Origin: A German nursery selection of a species wild in parts of the Middle East.

Height: 20 inches.

Spread: 15 inches.

Description: Pink flowers surrounded by red-purple bracts. Foliage with the distinctive herbal fragrance of oregano.

Season: July to September.

Garden Habitat and Cultivation: Full sun, any reasonable soil, including poor ones. This is one of the best later-flowering, drought-tolerant plants. "Should cut back in late spring so plants do not become too leggy," advises Jennifer, adding that "this also helps encourage blooms later in the summer, so it blooms at the same time as the sea lavender, making a gorgeous display."

Hardiness Zone—Minimum: 5.

Planting Companions: *Coreopsis verticillata* 'Golden Showers', *Dalea purpurea*, *Eryngium bourgatii*, *Limonium platyphyllum*, *Perovskia atriplicifolia* 'Little Spire', *Sedum* spp., *Vernonia lettermanii* 'Iron Butterfly'

***Paeonia lactiflora* 'Jan Van Leeuwen'**

herbaceous peony

Origin: Nursery selection of a species originally from China.

Height: 24 inches.

Spread: 30 inches.

Description: "Another winner, amazing fall color, always healthy," says Colleen of this single pure-white peony with a central boss of golden stamens. Emerging spring foliage is a deep-red color, turning to handsome divided green leaves; in fall, these turn gold and orange. The chunky seed heads are also an attractive feature.

Season: May, plus fall color.

Garden Habitat and Cultivation: Full sun or light shade. Any reasonable and fertile well-drained soil. Peonies are slow to establish, but long-lived.

Hardiness Zone—Minimum: 3.

8.30. *Related to the oregano of the kitchen,* Origanum laevigatum *'Herrenhausen' is a good small later-flowering perennial for drier sites. It will often self-seed. (Photo: Piet Oudolf)*

Planting Companions: *Geranium*, *Hosta*, and *Hemerocallis* varieties. Small blue bulbs, such as species of *Muscari* in spring look spectacular against the young red leaves.

Papaver orientale 'Scarlet O'Hara'—Scarlet O'Hara Poppy (horticultural origin)

oriental poppy

Origin: A garden hybrid between species native to Turkey and the Middle East.

Height: 30 inches.

Spread: 20 inches.

Description: Very large scarlet flowers—classic poppy shape and hard to ignore! Big seed heads provide some interest after the rest of the plant dies back.

Season: June to July.

Garden Habitat and Cultivation: Full sun and any reasonable soil, some drought tolerance. The plants are adapted to spring/early summer growth and summer dormancy, so they will naturally die back by the end of July, leaving something of a gap, so they mesh well with late-developing perennials. They dislike high summer temperatures and humidity.

Hardiness Zone—Minimum: 3.

Planting Companions: Blue-mauve *Nepeta* species make a good color contrast, and the plants are matched in height and scale. *Digitalis ferruginea* and varieties of *Veronicastrum* a good contrast in form. In the Lurie Garden, it is used as a theme plant dotted through a large area to create a big-scale splash.

Parthenium integrifolium

wild quinine

Origin: United States east of the Great Plains.

Height: 36 inches.

Spread: 20 inches.

Description: Flat-topped clusters of white flower heads on upright stems, each one looking slightly as if it were made of melted plastic; these are

8.31. **Parthenium integrifolium,** ***wild quinine, is relatively new to cultivation. Not colorful, but it has good structure and, like many white or cream flowers, helps to blend other stronger colors. (Photo: Noel Kingsbury)***

very sturdy and, as Jennifer points out, "perfect for catching falling snow." Very popular with bees and butterflies.

Season: June to July.

Garden Habitat and Cultivation: Full sun and any reasonable soil, some drought tolerance.

Hardiness Zone—Minimum:4.

Planting Companions: *Amsonia* spp., *Asclepias incarnata*, *Astilbe chinensis* var. *taquetii* 'Purpurlanze', *Hemerocallis* varieties, *Scutellaria incana*.

Penstemon digitalis 'Husker Red'
foxglove beard tongue

Origin: Eastern Canada and eastern half of the United States.

Height: 24 inches.

Spread: 20 inches.

Description: A bold clump of basal leaves with a distinct dark-red color; stems of white tubular flowers.

Season: April to May.

Garden Habitat and Cultivation: Sun or light shade, any reasonable soil, some drought tolerance. This is a true herbaceous perennial, so it dies back completely in the winter—unlike the many western US species of *Penstemon*, which are almost shrubby in character and much less hardy.

Hardiness Zone—Minimum: 3.

Planting Companions: *Geranium* 'Brookside', *Geum rivale* 'Flames of Passion', *Nepeta* spp., *Salvia*—meadow salvias, *Sanguisorba menziesii*.

Perovskia atriplicifolia 'Little Spire'
Russian sage

Origin: A nursery selection of a species found across central Asia in habitats similar to North American sagebrush country. The conditions in which it thrives are unbelievably harsh.

Height: 24 inches.

Spread: 20 inches.

Description: Pretty and useful for the midsummer garden, with violet-blue flowers on more or less upright spikes. These continue to look good after flowering. White skeletal remains in the winter.

Season: June to September.

Garden Habitat and Cultivation: Full sun, any well-drained soil. Very drought tolerant. Probably quite happy in very stony soils.

Hardiness Zone—Minimum: 4.

Planting Companions: *Ceratostigma plumbaginoides*, *Eryngium yuccifolium*, *Knautia macedonica*, *Limonium platyphyllum*, *Nepeta* varieties, *Origanum laevigatum* 'Herrenhausen', *Salvia*—meadow sages, *Sedum* 'Matrona', *Vernonia lettermanii* 'Iron Butterfly'.

Persicaria amplexicaulis 'Firedance'
a species of knotweed

Origin: A nursery selection of a Himalayan species.

Height: 50 inches.

Spread: 45 inches, slowly but surely advancing to form large clumps.

Description: Crimson-red "tails" adorn an almost shrubby-looking

8.32. **Persicaria amplexicaulis *'Firedance', a variety of knotweed, is potentially a valuable mid- to late-summer-flowering perennial, but only for moister soils and cooler regions. (Photo: Noel Kingsbury)***

mound of large leaves over a long season. 'Alba' is a white form. There are also an increasing number of others available commercially, many with flowers in shades of pink.

Season: June to October.

Garden Habitat and Cultivation: Full sun or light shade; moist, fertile soils needed for a good performance. Potentially a superb garden plant but unfortunately pest prone (e.g., mites) and liable to rabbit damage.

Hardiness Zone—Minimum: 4.

Planting Companions: *Amorpha canescens*, *Aruncus* 'Horatio', *Aster* and *Sym-*

8.33. *The seed heads of* Phlomis tuberosa *'Amazone' can make a strong feature in the fall garden if they are not deadheaded. Like any seed heads that stand well into winter, they are a good food resource for seed-eating birds. (Photo: Robin Carlson)*

phyotrichum species, *Eurybia divaricata*, *Echinacea* spp., *Echinops bannaticus* 'Blue Glow', *Eryngium yuccifolium*, *Eupatoriadelphus maculatus*, *Geranium* 'Jolly Bee', *Geranium x oxonianum* 'Claridge Druce', *Inula magnifica* 'Sonnenstrahl', *Pycnanthemum muticum*, *Salvia azurea*.

Phlomis tuberosa 'Amazone'
tuberous Jerusalem sage

Origin: Nursery selection of a plant found across eastern Europe and parts of central Asia.
Height: 50 inches.
Spread: 20 inches.
Description: Upright stems carry whorls of purple-pink, beaked flowers. Dark, arrow-shaped leaves.

Season: Mid- to late June.

Garden Habitat and Cultivation: Full sun or light shade, any reasonable soil. Colleen notes that it is "so aggressive at seeding, it felt out of control at times, we'd even call the garden 'phlomisland,' but it did well under trees in the Dark Plate, which was one reason why we kept it." Despite this, she does note that "it is easy to manage by pulling out seedlings." "We just remove half to two-thirds of the seed heads after blooming," says Jennifer. "This leaves enough seed heads for fall and winter interest, but substantially reduces the amount of weeding."

Hardiness Zone—Minimum: 5.

Planting Companions: *Amsonia* spp., *Baptisia* spp., *Filipendula rubra* 'Venusta Magnifica', *Monarda bradburyana*, *Pycnanthemum muticum*, *Scutellaria incana*.

Polygonatum biflorum

Solomon's seal

Origin: Canada and United States east of the Rockies.

Height: 30 inches.

Spread: 18 inches, slowly forming extensive clumps.

Description: Arching stems carry pendant, green-tipped creamy bells, followed in fall by blue-black berries. Golden fall color.

Season: May for flowers, October for leaf color.

Garden Habitat and Cultivation: Shade, moist but well-drained soil. Can be slow to establish.

Hardiness Zone—Minimum: 3.

Planting Companions: *Epimedium* spp., *Helleborus x hybridus*, *Jeffersonia diphylla*, *Mertensia virginica*, *Zizia aurea*.

Polystichum setiferum 'Herrenhausen'

soft shield fern

Origin: Southern Europe.

Height: 18 inches.

Spread: 18 inches.

Description: Classic fern fronds radiating out from a central boss; in fact, these are among the finest and most elegant of any cultivated fern.

There are numerous similar varieties that differ in their level of frond division.

Season: Spring to autumn foliage.

Garden Habitat and Cultivation: Part to full shade, any reasonable soil. Although it grows best in a well-drained, moist soil, it shows a surprising level of tolerance of drought and of sunshine compared to other ferns. Avoid cold and exposed situations and waterlogged soils.

Hardiness Zone—Minimum: 5.

Planting Companions: *Epimedium* spp., *Geranium x cantabrigiense* 'Karmina', *Helleborus x hybridus, Jeffersonia diphylla, Mertensia virginica, Polygonatum biflorum.*

Pycnanthemum muticum
clustered mountain mint

Origin: Eastern United States.

Height: 36 inches.

Spread: 24 inches, spreading to form large clumps.

Description: Upright stems of intensely minty scented foliage and silvery bracts around the tiny pink flowers. Much loved by bees.

Season: July onward.

Garden Habitat and Cultivation: Full sun or light shade, moist soils preferred. Little drought tolerance—other species of *Pycnanthemum* are better on dry soils; the plant has also limited cold tolerance.

Hardiness Zone—Minimum: 5.

Planting Companions: *Amorpha canescens, Anemone*—Japanese types, *Asclepias incarnata, Echinops bannaticus* 'Blue Glow', *Hemerocallis* varieties, *Persicaria amplexicaulis* 'Firedance', *Phlomis tuberosa* 'Amazone', *Scutellaria incana.*

Rodgersia pinnata 'Superba'
featherleaf rodgersia

Origin: Southwestern China.

Height: 45 inches.

Spread: 40 inches.

Description: Huge lush-looking leaves, divided into several leaflets, bronze at first, turning dark green with a rough-looking surface, make a splendid foil for large heads composed of masses of tiny cream flowers. Among the most exotic-looking hardy plants for the Midwest.

Season: June to July for flowers, all the growing season for foliage.

Garden Habitat and Cultivation: Light shade. Moist, fertile soil is vital, especially in sunnier positions as the leaves scorch easily if too dry. The ideal position is by a streamside or on a slope where water is constantly moving through the soil. They can also be damaged by late frost.

Hardiness Zone—Minimum: 5.

Planting Companions: *Anemone leveillei*, *Aruncus* 'Horatio', *Astilbe chinensis* var. *taquetii* 'Purpurlanze', *Geranium x oxonianum* 'Claridge Druce', *Hosta* varieties, *Hemerocallis* varieties, *Lythrum alatum*, *Sanguisorba* spp.

Ruellia humilis

wild petunia

Origin: Midwestern and eastern United States.

Height: 24 inches.

Spread: 24 inches.

Description: Large, open, five-petaled lilac flowers; chunky, lobed leaves.

Season: Although starting into growth late (May), it flowers over a long period, July to fall.

Garden Habitat and Cultivation: Full sun to light shade. Any reasonable soil, with some drought tolerance. It can self-sow liberally, but the seedlings are slow to grow, so this is not a problem.

Hardiness Zone—Minimum: 4.

Planting Companions: *Coreopsis verticillata* 'Golden Showers', *Dalea purpurea*, *Echinacea* spp., *Hemerocallis* varieties, *Sedum* spp.

Salvia azurea

blue sage

Origin: Southeastern United States.

Height: 40–60 inches.

Spread: 24 inches.

Description: Few flowers are true blue, but this is one is, with pale-blue blooms that make a striking impact. However, it looks very undistinguished until it flowers and can have a rather floppy growth habit.

8.34. ***Wild petunia,* Ruellia humilis, *has a long season of flower, which, combined with its relatively small stature (less than 24 inches), makes it useful for bringing color to odd spaces in late summer. (Photo: Piet Oudolf)***

Season: September.

Garden Habitat and Cultivation: Full sun, any reasonable soil. Relatively drought tolerant.

Hardiness Zone—Minimum: 6.

Planting Companions: *Eupatoriadelphus maculatus*, *Persicaria amplexicaulis* 'Firedance', *Sanguisorba canadensis* 'Red Thunder', *Solidago rugosa* 'Fireworks', *Anemone*—Japanese types.

Salvia—meadow sages

The sages (*Salvia* species) are a huge group—botanists are undecided, but somewhere between seven hundred and nine hundred species. They reach their greatest diversity in the long chain of mountains between California and Argentina—few of these are reliably hardy, however. But a number of closely related European species (the meadow sages) are from a cold and seasonally dry continental habitat—comparable with shortgrass prairie. Nurseries have made hybrids involving at least three of these, and it makes sense to deal with them all together.

In the Lurie Garden, a number of meadow sage varieties have been used to create the Salvia River, which is such a distinct feature in May (see pages 10, 18).

Origin: Europe, parts of west Asia.

Height: 12 to 20 inches.

Spread: 12 inches.

Description: Bushy perennials with small flowers in plentifully produced spikes.

The following are in shades of blue violet:

S. nemorosa 'Amethyst', which seems to be the strongest grower of them all in the Lurie Garden, *S. nemorosa* 'East Friesland' and *S. nemorosa* 'Wesuwe', which according to Colleen is the most vivid, *S. x sylvestris* 'Blue Hill'; *S. x sylvestris* 'Dear Anja', *S. x sylvestris* 'May Night' and *S. x sylvestris* 'Rügen'.

There are a few pink varieties available, such as *Salvia pratensis* 'Pink Delight'.

Season: May, but most will repeat flower in September if the early flowers are deadheaded and the plants lightly pruned in June.

Garden Habitat and Cultivation: Full sun, any reasonable soil. Tolerant of drought and alkaline soils, but little tolerance of waterlogging. The plants do not spread—but some self-seed in some gardens. The plants are reasonably long-lived but do not survive forever, especially in regions with high summer humidity.

Hardiness Zone—Minimum: 4.

Planting Companions: *Asclepias tuberosa*, *Coreopsis verticillata* 'Golden Showers', *Dalea purpurea*, *Eryngium bourgatii*, *Geranium sanguineum* 'Max Frei',

Geum rivale 'Flames of Passion', *Geum triflorum*, *Perovskia atriplicifolia* 'Little Spire', *Stachys officinalis*, *Vernonia lettermanii* 'Iron Butterfly', *Zizia aurea*.

Sanguisorba canadensis 'Red Thunder'
Canadian burnet

Origin: Eastern North America.

Height: 75 inches.

Spread: 30 inches.

Description: A rather majestic plant with upright stems carrying fluffy-looking spikes of tiny burgundy-red flowers. The divided leaves are elegant from the moment they emerge in spring. Good yellow fall color.

Season: August to September.

Garden Habitat and Cultivation: Full sun, any reasonable soil, but little tolerance of drought.

Hardiness Zone—Minimum: 3.

Planting Companions: *Amorpha canescens*, *Symphyotrichum novae-angliae*, *Aster* 'Blue Autumn', *Symphyotrichum oblongifolius* 'October Skies', *Inula magnifica* 'Sonnenstrahl', *Salvia azurea*, *Scutellaria incana*.

Sanguisorba menziesii
a species of burnet

Origin: Alaska and Pacific Northwest.

Height: 30 inches.

Spread: 18 inches.

Description: Leaves divided into multiple, blue-green leaflets. Pinky-red, short flower spikes that remain a good, dark red-brown color even as they fade into seed heads.

Season: May–June.

Garden Habitat and Cultivation: Sun and any reasonable soil, but intolerant of drought. Can look untidy after flowering but a haircut encourages attractive fresh young growth.

Hardiness Zone—Minimum: 3.

Planting Companions: *Amsonia* spp., *Anemone leveillei*, *Astilbe chinensis* var. *taquetii* 'Purpurlanze', *Baptisia* spp., *Geranium* 'Brookside', *Geranium* x *cantabrigiense* 'Karmina', *Gillenia trifoliata*, *Monarda bradburyana*, *Nepeta racemosa* 'Walker's Low', *Stachys officinalis*.

Scutellaria incana
hoary skullcap
Origin: Northeastern United States.
Height: 36 inches.
Spread: 20 inches, slowly spreading.
Description: Delicate pale-blue flowers in a rather intricate shape gathered on loose spikes—"baseball cap" would actually be a more accurate name than "skullcap." These are followed by spectacular seed heads. The leaves have a distinctly hairy underside that appears white. Good fall color—red, purple, and bronze tones. "Great, great plant," says Colleen.
Season: July to September.
Garden Habitat and Cultivation: Full sun or light shade, any reasonable soil, some drought tolerance. The plants slowly form clumps and often self-seed, but not excessively.
Hardiness Zone—Minimum: 5.
Planting Companions: *Agastache rupestris*, *Asclepias incarnata*, *Echinacea* spp., *Lythrum alatum*, *Parthenium integrifolium*, *Phlomis tuberosa* 'Amazone', *Pycnanthemum muticum*, *Sanguisorba canadensis* 'Red Thunder'.

Sedum x hybrida 'Bertram Anderson'
a variety of stonecrop
Origin: Garden origin, parents from Japan and Europe.
Height: 8 inches.
Spread: 15 inches.
Description: Very dark reddish-gray leaves form a small mound of great distinctiveness; purple-pink flower clusters.
Season: August to September.
Garden Habitat and Cultivation: Full sun, any reasonable soil, some drought tolerance. Wet soils or overhead irrigation can lead to rotting—perhaps advisable only on really well-drained soils.
Hardiness Zone—Minimum: 5.
Planting Companions: *Asclepias tuberosa*, *Coreopsis verticillata* 'Golden Showers', *Gentiana andrewsii*, *Geranium* 'Jolly Bee', *Limonium platyphyllum*,

8.35. ***This stonecrop variety (*Sedum *'Matrona') is drought tolerant and long-lived, with seed head interest until crushed by snow. Most of the stonecrops are very good for attracting butterflies. (Photo: Piet Oudolf)***

Origanum laevigatum 'Herrenhausen', *Perovskia atriplicifolia* 'Little Spire', *Ruellia humilis*, *Vernonia lettermanii* 'Iron Butterfly'.

Sedum 'Matrona' (Hylotelephium 'Matrona')
a variety of stonecrop

Origin: A hybrid from a German nursery, between two Eurasian species.
Height: 30 inches.
Spread: 24 inches.
Description: Masses of pale-pink flowers in flattish heads on a plant with a more upright habit than many other sedums, grayish slightly succulent leaves.
Season: August to September.
Garden Habitat and Cultivation: Full sun, any reasonable soil, plus relatively poor ones, good drainage important, drought tolerant.
Hardiness Zone—Minimum: 3.
Planting Companions: As above.

8.36. **Silphium laciniatum** *(compass plant) is one of several long-lived prairie species known as prairie docks, but it is probably the most striking, for its large deeply divided leaves as much as for its immensely tall flower stems. In fall and early winter, these leaves can be particularly dramatic if backlit by the sun. (Photo: Piet Oudolf)*

Silphium laciniatum
compass plant

Origin: Central and northeastern US states, southeast Canada.

Height: Can grow to 9 feet! This is mostly flower stalk, however.

Spread: 30 inches.

Description: Big (to 5 inches) yellow daisies top the giant stems. It is the leaves, however, that are really extraordinary and quite unlike anything else in the daisy family, or indeed anything else in the North American flora: around 18 inches long, deeply divided, and very hard and leathery—a bit like giant oak leaves. They also orientate themselves on a north-south axis—hence the common name. They are spectacular throughout the summer and into fall, being particularly dramatic when backlit by the sun.

Season: August.

Garden Habitat and Cultivation: Full sun, any fertile soil. They are among several classic prairie species that spend their first few years building

up a root system (which may go down 20 feet), so they may be slow to flower. They do not form clumps but can self-seed.
Hardiness Zone—Minimum: 3.
Planting Companions: "Dynamite in the meadow," says Colleen. This tall and dramatic plant is liable to look out of place among smaller plants, but if grown among other tall perennials, its leaves may be concealed. Probably best grown as part of plant mix that evokes its natural prairie habitat with medium-sized grasses and perennials.

Solidago rugosa 'Fireworks'
rough, wrinkleleaf goldenrod
Origin: A selection made at North Carolina Botanic Gardens. The species is found across eastern North America.
Height: 36 inches.
Spread: 24 inches, steadily forms a clump, but not aggressively.
Description: Goldenrods suffer from the "familiarity breeds contempt" syndrome—this, however, is the most elegant. Upright stems with small, rough leaves and flower heads that appear to zigzag through the air like fireworks. This species has been joined at the Lurie by a recently introduced cultivar, 'Wichita Mountains'; it has stubby cylindrical heads of yellow flowers. Found in southwestern Oklahoma, it is much more heat and drought tolerant than other goldenrods.
Season: September–October.
Garden Habitat and Cultivation: Full sun; any reasonable soil, including wet ones. May self-seed.
Hardiness Zone—Minimum: 4.
Planting Companions: *Aster tataricus* 'Jindai', *Gillenia trifoliata*, *Salvia azurea*, *Symphyotrichum novae-angliae*, *S.* 'Blue Autumn', *S. oblongifolius* 'October Skies'.

Stachys officinalis
betony, hedgenettle
Origin: Europe, North Africa, western Asia.
Height: 24 inches.
Spread: 18 inches.
Description: Modestly sized but prolific rounded spikes of small flowers.

8.37. **Solidago rugosa** ***'Fireworks' has clusters of yellow flowers seeming to zigzag off in all directions. Like many other late-flowering members of the daisy family, it is a very good butterfly plant. (Photo: Noel Kingsbury)***

Foliage is nondescript at a distance, but on closer examination each leaf resembles that of a tiny oakleaf. There are several varieties in cultivation—the original species is an intense reddish pink, 'Rosea' is pale pink, and 'Hummelo' is a stronger pink and a stronger plant.

Season: May to June.

Garden Habitat and Cultivation: Full sun and any reasonable soil.

Hardiness Zone—Minimum: 4.

Planting Companions: *Amsonia* spp., *Dalea purpurea*, *Geranium x oxonianum*

'Claridge Druce', *Geranium sanguineum* 'Max Frei', *Gillenia trifoliata*, *Liatris spicata*, *Monarda bradburyana*, *Nepeta* spp., *Salvia*—meadow sages, *Sanguisorba menziesii*.

Symphyotrichum novae-angliae 'Violetta' (_Aster novae-angliae_ 'Violetta')
New England aster

Origin: A nursery selection of a species common to North America east of the Rockies.

Height: 72 inches.

Spread: 30 inches.

Description: One of the most familiar asters, even to anyone whose experience of wildflowers is limited to what can be seen on the verges of interstate highways; such a person might even notice, as they come off and on the ramp, just how much natural variation in color there is. 'Violetta' is simply one of many garden varieties, which can range in color from white through pink to pink red in one direction and through mauve blue to violet in another.

Season: Late August to September.

Garden Habitat and Cultivation: Full sun and average soil. A very tolerant and easy plant, with one big problem—rabbits love it!

Hardiness Zone—Minimum: 4.

Planting Companions: *Amorpha canescens*, *Symphyotrichum* 'Blue Autumn', *S. oblongifolius* 'October Skies', *Aster tataricus* 'Jindai', *Eupatoriadelphus maculatus*, *Persicaria amplexicaulis*, *Sanguisorba canadensis* 'Red Thunder', *Solidago rugosa* 'Fireworks'.

Symphyotrichum 'Blue Autumn' (_Aster_ 'Blue Autumn')
a variety of aster

Origin: Nursery bred in Holland, ancestors North American.

Height: 20 inches.

Spread: 16 inches.

Description: Masses of deep purple-blue daisy flowers with yellow centers on a compact plant with healthy foliage. A plant for making a show!

Season: September to October.

Garden Habitat and Cultivation: Full sun, light shade, any reasonable soil. Unlike some asters, it is relatively mildew resistant.

8.38. Symphyotrichum *'Blue Autumn' is one of many late-flowering asters with good blue flowers. Here, it is seen with* Calamintha nepeta *subsp.* nepeta. *Other good color combinations would be with pink-flowered asters or some of the many yellow-flowering natives, such as species of* Helianthus, Rudbeckia, *and* Solidago. *(Photo: Robin Carlson)*

Hardiness Zone—Minimum: 3.

Planting Companions: *Symphyotrichum oblongifolius* 'October Skies', *Aster tataricus* 'Jindai', *Coreopsis verticillata* 'Golden Showers', *Persicaria amplexicaulis, Sanguisorba canadensis* 'Red Thunder', *Solidago rugosa* 'Fireworks'.

Symphyotrichum oblongifolius 'October Skies' (_Aster oblongifolius_ 'October Skies')

aromatic aster

Origin: A nursery selection of a plant native to the United States east of the Rockies.

Height: 18 inches.

Spread: 24 inches.

Description: Masses of lavender-blue flowers on a low and mounding multiple-branched plant—very different to the upright habit of most asters.

Season: September to October.

Garden Habitat and Cultivation: Sun or light shade, any reasonable soil; successful on infertile soil, some drought tolerance.

Hardiness Zone—Minimum: 4.

Planting Companions: *Symphyotrichum* 'Blue Autumn', *Coreopsis verticillata*

'Golden Showers', *Aster tataricus* 'Jindai', *Eupatoriadelphus maculatus*, *Persicaria amplexicaulis*, *Sanguisorba canadensis* 'Red Thunder', *Solidago rugosa* 'Fireworks'. Because of its habit, one of the few asters that can be used for groundcover.

Thalictrum 'Elin'
a species of meadow rue

Origin: A nursery hybrid.
Height: 70 inches.
Spread: 30 inches.
Description: Heads of tiny cream flowers held aloft on upright stems with purple-flushed foliage. Magnificent.
Season: June.
Garden Habitat and Cultivation: Sun or light shade, any reasonable soil, but little tolerance of drought. May be intolerant of summer heat farther south.
Hardiness Zone—Minimum: 4.
Planting Companions: This plant is so much taller than anything else that flowers in early summer it is difficult to suggest "companions" as such—possibilities include lower-level plants with a similar liking for moister soils such as *Geranium* or use it for bringing early summer color to combinations of later-flowering prairie species.

Tricyrtis
toad lilies

Intriguing late-flowering plants originally from Japan, distinct for their heavily spotted flowers; an increasing number of hybrids and varieties are becoming available.
Height: Varieties used in the Lurie are around 30 inches.
Spread: 18 inches.
Description: *Tricyrtis formosana* has lavender flowers with dark-purple spotting. T. 'Tojen' has lilac flowers with a yellow center, minimal spotting but very free flowering.
Season: August–September.
Garden Habitat and Cultivation: Light to full shade with moist, humus-rich soil that never dries out in summer is essential. Mulching and

irrigation will help survival—Japan has very wet summers. *T. formosana* can form large clumps with time.
Hardiness Zone—Minimum: 6.
Planting Companions: *Anemone*—Japanese types, *Gentiana andrewsii*, *Geranium* 'Jolly Bee'. Although not flowering at the same time, of course, spring-blooming woodlanders like *Jeffersonia diphylla* and *Polygonatum biflorum* are habitat companions.

Vernonia lettermanii 'Iron Butterfly'
narrowleaf ironweed
Origin: A selection of a species native to the Ozark Mountains.
Height: 24 inches.
Spread: 24 inches.
Description: Small purple heads, like diminutive thistle heads. A very good plant for attracting butterflies—as are all ironweeds.
Season: September to first frosts.
Garden Habitat and Cultivation: Any well-drained soil, including dry and stony ones.
Hardiness Zone—Minimum: 4.
Planting Companions: *Asclepias tuberosa*, *Geranium sanguineum* 'Max Frei', *Knautia macedonica*, *Limonium platyphyllum*, *Nepeta* spp., *Origanum laevigatum* 'Herrenhausen', *Salvia*—meadow sages.

Veronica longifolia 'Eveline'
a variety of speedwell
Origin: A nursery selection of a north Asian species.
Height: 30 inches.
Spread: 12 inches.
Description: Purple-blue flowers in densely packed, upright spikes.
Season: May to July.
Garden Habitat and Cultivation: Full sun and fertile soils, preferably not drying out. Other varieties of *Veronica longifolia* have been used in the Lurie Garden but have not performed well—they are eaten by rabbits, but other factors such as high summer temperatures may be a problem.
Hardiness Zone—Minimum: 4.
Planting Companions: Varieties of *Geranium*.

8.39. ***Increasingly popular in gardens,*** **Veronicastrum virginicum,** ***known traditionally as Culver's root, is being brought out in more and more varieties, which differ in color, degree of branching, and habit. All, however, are distinctly upright and flower in mid- to late summer with good ongoing winter structure. This is 'Rosea'. (Photo: Piet Oudolf)***

Veronicastrum virginicum
Culver's root
Origin: Eastern half of North America.
Height: 75 inches.
Spread: 30 inches.
Description: The wild species has tiny, pale lavender-mauve flowers packed onto narrow spikes, usually branching. The plant is as valuable for its form as anything else—with stiff, upright stems and neat whorls of narrow leaves, it is elegant and structural at all times. 'Temptation' has more strongly colored purple flowers and is somewhat shorter than the species. 'Pink Glow' has very pale-pink, almost white flowers, 'Rosea' has pink flowers, 'Diane' has white flowers, a little later than other varieties.

Season: June flowering, fall seed heads.

Garden Habitat and Cultivation: Full sun and fertile soils, preferably not drying out.

Hardiness Zone—Minimum: 3.

Planting Companions: *Anemone*—Japanese types, *Asclepias incarnata*, *Echinops bannaticus* 'Blue Glow', *Eryngium yuccifolium*, *Eupatoriadelphus maculatus*, *Filipendula rubra* 'Venusta Magnifica', *Geranium x oxonianum* 'Claridge Druce'.

Zizia aurea

golden alexanders, golden zizia

Origin: Central and eastern Canada and United States.

Height: 36 inches.

Spread: 20 inches. Forms a tight nonspreading clump.

Description: Rounded heads of yellow flowers with a green tinge over glossy green leaves.

Season: May.

Garden Habitat and Cultivation: Sun or light shade. Self-sows.

Hardiness Zone—Minimum: 3.

Planting Companions: Blue *Amsonia* and *Camassia* species are the obvious companions, and *Salvia*—meadow types. The greeny yellow makes a wonderful combination with tulips. Colleen's comment on it is that it "blended so well with purple, adjacent to the Salvia River, and *Amsonia* willowleaf—all looked so good together."

GRASSES

Calamagrostis brachytricha

Korean feather reed grass

Origin: Central and eastern Asia.

Height: 45 inches.

Spread: 18 inches.

Description: An upright grass with fine flower/seed heads of an elongated shape and moderately broad fresh green leaves. When in flower, the heads are pinkish purplish, but fade to a grayish tone as they seed. It becomes completely dormant in winter.

Season: August to November.

8.40. ***Larger ornamental grasses such as* Calamagrostis *'Karl Foerster' can make a dramatic impact in the later-season garden. This particular variety stands strongly from its June flowering well into the winter, its color appearing to change with different light conditions—it can be relatively dark, to glowing chestnut brown, to bleached straw. (Photo: Noel Kingsbury)***

Garden Habitat and Cultivation: Full sun, any reasonable soil, plus moderately wet ones. Dislikes competition. Can seed aggressively.

Hardiness Zone—Minimum: 4.

Planting Companions: Ideal for mixing with lower-growing, early flowering perennials to create interest for later; looks good alongside summer-flowering perennials of similar height.

Calamagrostis x acutiflora **'Karl Foerster'** (*C. x acutiflora* **'Stricta'**)
feather reed grass

Origin: A hybrid between two Eurasian species, grown by the great German nurseryman and writer of the early twentieth century Karl Foerster.

Height: 60 inches.

Spread: 30 inches, slowly but surely spreading to form large clumps.

Description: Simply one of the most valuable ornamental grasses, with firm, upright stems, making a perfect punctuation mark for the garden. Heads have an attractive habit of moving in the breeze. Leaves stay green in milder winters.

Season: June to midwinter.

Garden Habitat and Cultivation: Full sun, any reasonable soil and wet ones. Used by the leading landscape designer on the East Coast, James van Sweden, as it is hurricane proof—"it just bends over and bounces back up," he says. Cut back in late winter or whenever it begins to look too scruffy. It does not run and is sterile so does not seed.

Hardiness Zone—Minimum: 5.

Planting Companions: Just about any perennials that are noticeably shorter, although it can be used to create a stylized prairie look with plants of similar size. Very useful for creating a sense of repetition and rhythm in the garden.

Carex muskingumensis
palm sedge

Origin: A band between Oklahoma and Ontario.

Height: 36 inches.

Spread: 24 inches, spreading slowly to form a clump.

Description: Even nongardeners tend to see that this sedge offers something different from the usual "sit on the ground and grow a clump of green leaves" sedgey look. This one produces a mound of weakly upright stems with a succession of midgreen, slightly shiny leaves. Moderately interesting yellow fall color.

Season: May to the first frosts, as a foliage plant.

Garden Habitat and Cultivation: Sun or light shade, any reasonable soil and wet ones. Some drought tolerance.

8.41. ***Humble, but increasingly valued in gardens for their tolerance of poor soils and ground-covering ability, sedges come in many shapes and sizes. This one,*** **Carex pennsylvanica,** ***can be used as a lawn substitute in some situations. (Photo: Jennifer Davit)***

Hardiness Zone—Minimum: 4.

Planting Companions: More or less anything of comparable size. Effective with ferns and hostas in damp shade.

Carex pennsylvanica

Pennsylvania sedge

Origin: Eastern half of Canada and the United States.

Height: 10 inches.

Spread: 12 inches, continually spreading.

Description: Fresh green fine leaves and a running habit have persuaded some to use this as a lawn grass substitute—which is effective, but it will not take much walking on, let alone tennis or football.

Season: May to the first frosts; in mild winter climates, it can be almost evergreen.

Garden Habitat and Cultivation: Sun (if in moist soil) or shade, any soil including poor and dry ones. Best used as a lawn substitute in places where grass grows poorly, such as under trees, or on poor sandy soils—however, it cannot be walked on like grass.

Hardiness Zone—Minimum: 4.

Planting Companions: Vigorous ground-cover perennials such as the smaller *Geranium*, or larger shade lovers like *Hosta*.

Chasmanthium latifolium (Uniola latifolia)

northern sea oats

Origin: Eastern United States, plus Texas to Arizona.

Height: 60 inches.

Spread: 30 inches, slowly forming large clumps.

Description: A grass with broad leaves of fresh green and very distinctive pendant flower/seed heads that look a bit like oats that have been flattened with an iron—these catch the light nicely and move delicately in the breeze.

Season: August to September.

Garden Habitat and Cultivation: Sun or light shade, any reasonable soil and moderately wet ones.

Hardiness Zone—Minimum: 3.

Planting Companions: Shade-tolerant perennials, or scatter among spring-flowering perennials and bulbs in shade for later-season interest.

Eragrostis spectabilis

purple love grass

Origin: Eastern Canada, United States east of the Rockies, Mexico.

Height: 24 inches.

Spread: 24 inches.

Description: Clouds of tiny purple-brown flowers hover over clumps of moderately broad leaves.

Season: July to September.

Garden Habitat and Cultivation: Sun, most soils including poorer ones, drought tolerant. Short-lived but often self-sows—sometimes to problem proportions.

8.42. ***Purple love grass,*** **Eragrostis spectabilis,** ***has finely textured flowers and stems whose softness is a good contrast to more defined plant shapes. (Photo: Robin Carlson)***

Hardiness Zone—Minimum: 5.

Planting Companions: Flowering perennials of comparable size, or other grasses with contrasting and stronger structure.

***Miscanthus sinensis* 'Malepartus'**

common eulalia grass

Origin: A German-bred variety of a species native to eastern Asia.

Height: 90 inches.

Spread: 40 inches, slowly forming extensive clumps.

Description: A large, robust-looking grass with red-brown flowers that turn to silver-gray seed heads, in fingered clusters, which all line up in the lee of the wind. A favorite subject of Japanese traditional art.

Season: August to late winter. Seed heads stand the winter well.

Garden Habitat and Cultivation: Full sun or very light shade, any reasonable soil. Many *Miscanthus* grasses have become invasive aliens in southern states—so care should be taken to prevent seeding beyond the garden boundary.

Hardiness Zone—Minimum: 5.

Planting Companions: Other big muscular perennials.

Molinia caerulea

moor grass

Origin: Northern Eurasia, northeastern United States, and eastern Canada.

Height: Depends on variety.

Spread: Allow one-third of height.

Description: Tight clumps of fine foliage send up straight stems with brown flowers and seed heads. Stems are generally a vivid yellow. 'Dauerstrahl' is one of the taller cultivars, at 48 inches, with relatively upright spikes. 'Moorflamme' is similar but smaller, at 40 inches. There are increasing numbers of other cultivars available.

Season: July to November.

Garden Habitat and Cultivation: Sun; tolerant of poor, acidic and badly drained soils, but possibly not of dry ones. By December, the flower stems snap off at the base—this may seem a problem, but it makes clearing up easier, as one garden designer once noted, "You don't

have to cut them, just pick them up and chuck them on the compost heap."
Hardiness Zone—Minimum: 4.
Planting Companions: Very effective in masses, or can be mixed with later-flowering perennials of comparable size. They are quite late into growth, which makes them good companions for bulbs and early flowering perennials.

Molinia caerulea subsp. _arundinacea_ 'Transparent'
tall moor grass
Origin: Central and southern Europe.
Height: 72 inches.
Spread: 30 inches.
Description: Taller than the species, these explode upward like fireworks, being very effective for late-summer impact. They manage the trick of being very large and not taking up much space at the same time—much of their thrust is upward and their flower/seed spikes are very airy. Their foliage is an arching mound of lush green, flower stems generally green, but becoming clear yellow in fall. "Definitely one of the Lurie's best grasses, more sophisticated," notes Colleen.
Season: July to November.
Garden Habitat and Cultivation: As above, but more tolerant of light shade.
Hardiness Zone—Minimum: 4.
Planting Companions: They look cramped if grown alongside other tall plants, better to use them among plants that are about half their height.

Panicum virgatum 'Shenandoah'
switchgrass, panic grass
Origin: A German selection of a species native to Canada and the United States east of the western coastal states.
Height: 48 inches.
Spread: 40 inches.
Description: Earlier generations of Midwest farmers would probably been amused by the thought of garden centers selling switchgrass—it is still one of the most widely distributed prairie grasses and a major part

8.43. ***Panic grass,* Panicum virgatum *'Shenandoah', has a striking red leaf coloration that starts in midsummer and gets deeper in fall. (Photo: Noel Kingsbury)***

of any grassland that establishes itself on wasteland. This selection has very good red autumn color, which in fact begins in July, and the enormous spray of tiny flowers/seeds that is typical of this species.

Season: August to late winter. Seed heads stand the winter well.

Garden Habitat and Cultivation: Full sun, light shade, any reasonable soil, preferring it moister.

Hardiness Zone—Minimum: 4.

Planting Companions: Large prairie perennials, *Amsonia* species for fall color, late-flowering perennials.

***Pennisetum alopecuroides* 'Cassian'**
fountain grass

Origin: A cultivar of a species native to eastern Asia.

Height: 30 inches.

Spread: 30 inches.

Description: It is the fine, fluffy, "you can't help but stroke me" look that gives *Pennisetum* grasses their garden space. This is a variable species;

8.44. ***A variety of little bluestem grass,*** **Schizachyrium scoparium** ***'The Blues', is a very important grass in the Lurie Garden, as it combines good fall colors with not being too tall. Common as a native species in many different open environments, it has great potential for nature-inspired garden plantings. However, this picture does illustrate one of its faults: it can flop—this is less likely to happen on poorer or drier soils. (Photo: Noel Kingsbury)***

this selection has light cream-colored flowers with the arching habit that is most typical of the species. Red foliage tints in the fall.
Season: August to December.
Garden Habitat and Cultivation: Full sun to light shade, most soils, some drought tolerance. Can self-sow.
Hardiness Zone—Minimum: 5.
Planting Companions: Flowering perennials of comparable height.

Schizachyrium scoparium 'The Blues'
little bluestem
Origin: A cultivar of a species widespread in North America.
Height: 40 inches.
Spread: 24 inches.
Description: One of the most common of all prairie grasses, and perhaps

justly one of the dominant grasses in the design of the Lurie Garden. This selection has striking blue-tinged foliage and the burgundy-red tones that so many admire in the species as they sweep by at 55 miles per hour on the freeway. Purple-bronze flower heads turn to silvery seed heads.

Season: August to October.

Garden Habitat and Cultivation: Full sun; most soils, especially poor ones; drought tolerant. May be short-lived or floppy on fertile or moist soils. Can be expected to self-sow.

Hardiness Zone—Minimum: 3.

Planting Companions: Good for combining with shorter late-flowering perennials.

Sesleria autumnalis

autumn moor grass

Origin: Southern Europe, Caucasus.

Height: 12 inches.

Spread: 12 inches, steadily forming mats.

Description: Not a grass to be grown for its flowers or seed heads, nor to walk across or lie on, but somewhere in-between, forming tight clumps of slightly yellowy green leaves, and not very conspicuous flowers in autumn. Best thought of as a ground cover for small areas.

Season: May to November, for foliage interest.

Garden Habitat and Cultivation: Full sun or light shade. Ideal for dry, alkaline soils, so potentially very useful. It dislikes wet soils, high humidity or prolonged high temperatures.

Hardiness Zone—Minimum: 5.

Planting Companions: Clearly, given their size, *Sesleria* grasses can only be used as companions for similarly sized plants. Their main value is that their pale-colored leaves make such an excellent foil for enhancing the colors of adjacent flowers.

***Sorghastrum nutans* 'Sioux Blue'**

Indian grass

Origin: North America east of the Continental Divide.

Height: 60 inches.

8.45. ***The leaves of autumn moor grass,*** **Sesleria autumnalis,** ***look particularly good when backlit. Here, it grows with*** **Erynium bourgatii** ***(foreground) and meadow salvias. (Photo: Robin Carlson)***

Spread: 30 inches.

Description: A major prairie grass, important for forage; but even without pet buffalo to feed, it often makes an excellent garden ornamental. This cultivar has been selected for its strongly upright growth and blue-colored leaves, turning yellow gold in fall; copper-colored flowers. Strong winter presence. In garden conditions, without the competition of other plants, it can flop over badly, especially on fertile soils; in the Lurie Garden it has been replaced by *Panicum virgatum* 'Heavy Metal'.

Season: August to late winter.

Garden Habitat and Cultivation: Full sun, most soils, some drought tolerance.

Hardiness Zone—Minimum: 4.

Planting Companions: Larger late-flowering perennials.

Sporobolus heterolepis

prairie dropseed

Origin: North America east of the Continental Divide.

Height: 35 inches.

Spread: 35 inches.

Description: A prairie grass of major importance in the Great Plains. Forming a clump of long, fine, almost hairlike leaves, the flower and seed heads rise up to form a cloudy mass, each one so small as to be clearly visible only close-up. Golden fall color, the persistent leaves turning light bronze brown in the winter. The foliage has a distinct smell, which most find pleasant, but some dislike.

Season: August to late winter.

Garden Habitat and Cultivation: Full sun, most soils including dry and stony ones; drought tolerant.

Hardiness Zone—Minimum: 3.

Planting Companions: A wonderful companion to many flowering perennials of a similar size, both to flowers and seed heads.

'Tara' is a more compact form, discovered by Roy Diblik.

BULBS

Allium

ornamental onions and garlics

This is a large group found across the Northern Hemisphere in a wide range of habitats, most of which are relatively modest plants. Some of the more colorful species have become popular garden plants, usually sold as bulbs. Among them, the most horticulturally important are the 'Drumstick' alliums, with flower clusters that are almost uniquely spherical and the advantage that they turn into good and long-lasting seed heads. They are plants that combine color and form to a degree that makes them irresistible to designers. The seed heads do not usually survive the winter well, but look good in dried flower arrangements.

A few 'Drumstick' species or hybrids have been around since the earlier part of the twentieth century, but recent years have seen an increase in the number of varieties, as breeders, mostly in the Netherlands, introduce more into the bulb trade. New varieties tend to have stronger colors or larger heads than older ones but are always much more expensive.

Garden Habitat/Cultivation for All Species: Full sun, and well-drained soil. Reasonably good summer drought tolerance.

Allium atropurpureum

Origin: Southeastern Europe into Turkey.

Height: 34 inches.

Spread: 6 inches.

Description: Less neatly spherical than some of the 'Drumstick' onions (see *A. hollandicum*), this one, in Colleen's words, "lifts the garden with its distinctive dark color; it's the plant which gets the most attention in June." The flowers are a dark red purple.

Season: June, seed heads through the rest of the summer.

Hardiness Zone—Minimum: 4.

Allium cernuum

nodding onion

Origin: Most of continental North America below the subarctic zone.

Height: 15 inches.

Spread: 6 inches.

Description: Heads of rounded, nodding flowers, elegant and quite different from other wild onions and garlics. Pink, but the color can vary from the insipid to the intense—so best to buy in flower! It has been suggested that this wild onion was the *chicagu* plant of Native American tribes in the Chicago area and the origin of the city's name; in fact, *chicagu* is the leafier *Allium tricoccum*, known as ramps. *Allium cernuum* has now been removed from the Lurie Garden as it is so invasive, but as an attractive regionally native plant, some gardeners may want to grow it.

Season: June to July.

Hardiness Zone—Minimum: 4.

Allium christophii
star of Persia
Origin: Iran north to Kazakhstan.
Height: 16 inches.
Spread: 12 inches.
Description: Huge round heads (8 inches across) of dull pink star-shaped flowers with an extraordinary metallic sheen. One of the most asked-about plants in the garden in May. The seed heads add interest to the garden for long after the flowers have finished, but are best picked and used for decoration inside before the winter sets in. Unusual and spectacular.
Season: May/June, seed heads stand well into the winter.
Hardiness Zone—Minimum: 5.

***Allium hollandicum* 'Purple Sensation' (*A. aflatunense* 'Purple Sensation')**
Origin: A garden hybrid of uncertain origin, ancestors from the Middle East and central Asia.
Height: 30 inches.
Spread: 6 inches.
Description: This bulb is the best known of the 'Drumstick' onions, pale violet with a 3-inch head. It is relatively inexpensive and so suitable for mass planting.
Season: May to early June, seed heads through the rest of the summer.
Garden Habitat/Cultivation: Full sun and well-drained soils important. Foliage dies down quickly after flowering, but plants generally flower again well the year after—they may not do this if the leaves are overshadowed or crowded by neighboring plants. Often this species, and indeed some other alliums, will self-seed to problem proportions.
Hardiness Zone—Minimum: 4.

Allium sphaerocephalon
roundheaded leek
Origin: Western Europe south to North Africa and east to Turkey.
Height: 36 inches.
Spread: 3 inches.

8.46. ***Long established in the bulb trade, this ornamental garlic,* Allium hollandicum *'Purple Sensation', is cheap enough to use in quantity, when it can create spectacular late-spring effects. (Photo: International Flower Bulb Centre)***

Description: This is one of those plants that is scarcely visible until it flowers, and then the dark red-purple flowers in a 1–2-inch-wide head make a real splash. Each plant has two extremely narrow, almost grasslike leaves. It needs to be grown in large numbers to work its magic; fortunately, the bulbs are cheap. Given its very slender nature, it is best combined with perennials as a "scatter plant."

Season: July to August.

Garden Habitat/Cultivation: Full sun, any well-drained soil. Self-seeding is possible and often desirable.

Hardiness Zone—Minimum: 1.

Allium 'Summer Beauty'

Origin: Roy Diblik found this in a garden outside of Madison, Wisconsin, in 1995. He identified it as a form of A. *angulosum* (a Siberian species) probably acquired from a bulb mail-order company many years before. However, it may be a hybrid.

8.47. ***Low growing and perfect for tidily edging borders,*** **Allium** ***'Summer Beauty' is early to midsummer flowering and spreads slowly. Bees seem to love it. (Photo: Noel Kingsbury)***

Height: 24 inches.

Spread: 12 inches.

Description: Of the vast number of ornamental onions and garlics, this one looks a winner. Each head of pale lilac-pink flowers does not have an enormous impact by itself, but so many are produced, and on such a neat plant, that the overall effect is very striking. The leaves appear very early in spring, and unlike many onions continue to hold until fall, when they turn yellow. The seed heads are sturdy enough to survive through much of the winter without too much damage from ice and snow.

Season: Late May to June. Winter seed heads.

Garden Habitat/Cultivation: Any site in full sun, with well-drained soil. Unlike many onions, this one is sterile, so self-seeding is not a problem. Over time the plants will form good-sized clumps.

Hardiness Zone—Minimum: 5.

8.48. ***This picture of* Anemone blanda *'Blue Shades' illustrates what can be a very successful if unconventional use of small bulbs—combining with grass under trees for spring color. They will flower and then die back quickly, so by June the grass can be cut again. (Photo: International Flower Bulb Centre)***

Anemone blanda 'Blue Shades'
Grecian windflower

Origin: Species originally from eastern Mediterranean region.
Height: 6 inches.
Spread: 6 inches.
Description: Mauve-blue flowers about an inch across. They look like daisies but are totally unrelated.
Season: April.
Garden Habitat/Cultivation: Ideal for places below deciduous trees, or among summer-flowering perennials, as they flower early, with the foliage dying back in early summer. They are naturally woodland plants,

so a humus-rich soil is appreciated. Can also be naturalized in light grass in shade.

Hardiness Zone—Minimum: 5.

Camassia

camas

Camases are a group of six bulbous species, all native to North America. They are naturally blue, although there are also cream-white garden forms. All have straplike leaves and emerge quickly in spring, dying back by midsummer. Typically, they can form large colonies, quite spectacular in flower. In the garden, they are useful for naturalizing in bulk in prairie- or meadow-type grassland; with time, the bulbs will form clumps as they produce offset bulbs, as well as sometimes spreading through seeding. They are generally bought as dry bulbs from specialist bulb suppliers.

Camassia cusickii

Cusick's camas, Indian blue hyacinth

Origin: Northwestern United States, southwestern Canada.

Height: 30 inches.

Spread: 10 inches.

Description: Pyramidal heads of pale-blue star-shaped flowers, magnificent in large groups.

Season: May.

Garden Habitat/Cultivation: Full sun, moist soil, but surprisingly drought tolerant.

Hardiness Zone—Minimum: 4.

***Camassia leichtlinii* 'Blue Danube'**

camas, quamash.

Origin: A nursery selection of a western US native.

Height: 40 inches.

Spread: 10 inches.

Description: Dark-blue flowers in a head above an erect stem.

Season: May.

8.49. **Camassia leichtlinii** ***'Blue Danube' is one of several varieties of camas available from bulb companies. All are long-lived bulbs for sunny places and can be naturalized in meadow plantings, where grass is not cut until after the leaves have died back in midsummer. (Photo: International Flower Bulb Centre)***

Garden Habitat/Cultivation: Full sun or light shade, moist soils, but surprisingly drought tolerant.
Hardiness Zone—Minimum: 5.

Chionodoxa forbesii 'Blue Giant'
glory of the snow
Origin: A nursery selection of a Turkish plant.
Height: 9 inches.
Spread: 6 inches.
Description: Bulb with intense blue star-shaped flowers with a central white eye. *Chionodoxa luciliae* 'Violet Beauty' is a related variety, *C. sardensis* has intensely deep-blue flowers with a white eye. There are also several others available from bulb specialists.

8.50. *Known as glory of the snow, varieties of* Chionodoxa *are valued for their cheerful splashes of blue across the bare ground of the garden as they emerge from the chill of winter. (Photo: International Flower Bulb Centre)*

Season: April.

Garden Habitat/Cultivation: Light shade, ideal for underplanting deciduous trees or shrubs. Most reasonable soils with good drainage. With time, and especially if the ground is not too frequently disturbed, these little bulbs can spread to form colonies. Chionodoxas are among the bulbs that can also be naturalized in light grass in shade. Jennifer notes that "our *Chionodoxa* display makes a stunning carpet of blue in early spring—it's a fabulous alternative to crocus—longer lasting with brighter flowers."

Hardiness Zone—Minimum: 3.

Crocus tommasinianus

crocus

Origin: Southeastern Europe.

Height: 6 inches.

Spread: 4 inches.

Description: Chunky goblets above short-lived, very narrow leaves. 'Barr's Purple' is violet purple, 'Ruby Giant' is a rich red purple.

Season: March.

Garden Habitat/Cultivation: Sun or light shade. Crocus are very ephemeral—appearing very early, but often lasting only a short period; by May, their leaves are already dying back. They are very prone to being dug up by rodents, as unlike many bulbs they have no toxins or nasty-tasting chemicals to protect them—which is how many bulbs ward off hungry wildlife.

Hardiness Zone—Minimum: 3.

Fritillaria pallidiflora

a fritillary species

Origin: Northwest China.

Height: 30 inches.

Spread: 9 inches.

Description: Broad, rather square, pale-yellow flowers hanging from the top of upright stems, with glaucous leaves. Unusual and distinctive. The foliage dies back by June.

Season: April to May.

8.51. **Muscari armeniacum,** ***one of the small bulbs known as grape hyacinths, are very tolerant of poor soils and a range of different conditions. (Photo: International Flower Bulb Centre)***

Garden Habitat/Cultivation: Unlike many bulbs, this one definitely prefers light shade and does not appreciate being baked in the summer sun; humus-rich soil.
Hardiness Zone—Minimum: 3.

Muscari armeniacum **'Superstar'**
grape hyacinth

Origin: A nursery selection of a bulb from the eastern Mediterranean region and the Caucasus.
Height: 8 inches.
Spread: 4 inches.
Description: Deep dark-blue spherical flowers packed onto spikes, with a distinct fragrance (which can only be realistically appreciated once picked). Linear foliage. In the Lurie Garden, this is used along the Salvia River to provide an early effect of massed blue before the salvias flower.
Season: April to May.
Garden Habitat/Cultivation: Full sun or light shade from deciduous trees

and any reasonable well-drained soil. Foliage may emerge very early in areas with short or mild winters; it can always be expected to die back by midsummer. Unlike many bulbs, it bulks up quickly and also self-seeds, so dramatic sheets of blue can be built up over the years.
Hardiness Zone—Minimum: 4.

Narcissus
daffodils
Daffodils are the quintessential spring flower. Unlike tulips and hyacinths, however, they repeat flower from year to year reliably across a wide range of climate zones—so once you have them in your garden, you should have them forever. Cultivated daffodils are derived from several species of European origin. Height varies considerably; tall varieties (the most common) are useful in large gardens, but short varieties tend to be far more elegant and clearly suited to smaller spaces. Vigor chiefly affects the rate at which the variety will increase the size of its clump over time. Flowering time is important for visual effects but is less important in a continental climate where flowering times are more concentrated.

A site in full sun or the light shade of deciduous trees and any reasonably fertile well-drained soil are all that are needed for success . . . and one other thing—that their foliage is not cut off before it dies back naturally; daffodils must be able to replenish their bulbs with nutrients in order to flower again next year. Daffodils are ideal for combining with perennials, especially since growing perennial clumps will do a great job of hiding the increasingly scruffy foliage during the early summer period.
Hardiness Zone—Minimum: Generally zone 3 or 4.

'Actaea'

Wide, flat white flowers and a very small orange cup, late April or May flowering. 15 inches. More tolerant of wet soils than most.

'Jenny'

Yellow, with reflexed petals, April. 12 inches.

'Lemon Drops'

Pale-yellow flowers, April. 12 inches.

8.52. **Narcissus** ***'Lemon Drops' is a medium-height daffodil, with delicately colored flowers. Varieties like this are becoming increasingly popular, with many gardeners preferring them to the more widely planted large trumpet flowers, especially for small gardens or containers. (Photo: Robin Carlson)***

'Thalia'

White, flowering March to April. 15 inches.

Scilla mischtschenkoana

Tubergen squill

Origin: Iran, Caucasus, southern Russia.

Height: 6 inches.

Spread: 6 inches.

Description: Ice-blue flowers followed by straplike leaves that persist for several months before they go dormant. A stunning color.

Season: One of the first to flower in the Lurie Garden from late March to April.

8.53. Scilla mischtschenkoana *with* Crocus vernus *'Flower Record'. These are among the many small bulbs that can do so much to bring life to the spring garden and that should continue to grow and spread in years to come. (Photo: International Flower Bulb Centre)*

Garden Habitat/Cultivation: Sun or light shade. Any reasonable soil but not wet. Over time, the plants will naturalize as each bulb will form a clump; they often self-sow too.
Hardiness Zone—Minimum: 2.

Tulipa
tulips

Spring without tulips is now unthinkable. These vividly colorful flowers have long been popular as cut flowers, for displays in containers, for garden borders, and in massed plantings in public spaces; with more good varieties becoming available every year, the march of the tulip seems unstoppable. There is, however, a "but"—unlike daffodils, crocuses, and many other bulbs, they cannot be guaranteed to flower again next year. The reason for this lies in their origins—as the garden hybrids are derived from a number of species from the Middle East and central Asia, which have dry climates where the bulbs get a good summer "baking" after they have died back in early to midsummer.

The tulips grown in the Lurie Garden, however, have been selected by a Dutch bulb expert, Jacqueline van der Kloet, who has made a special study of tulip longevity. Colleen reports that most do indeed flower well for three years at least.

Full sun is essential, together with fertile well-drained soil. Tulips will repeat flower best if there is not too much competition or shading from other plants around them, even once the bulbs have died back. It is also important that their foliage is not cut back after flowering.
Hardiness Zone—Minimum: 3.

Tulipa 'Ballade'

Elegant shape with dusky purple flowers edged in white. Flowering late April to early May. 18 inches.

Tulipa 'Don Quichotte'

Rose-pink flowers in early May. 20 inches.

Tulipa 'Ivory Floradale'

Ivory flowers in a classic tulip shape, late April. 22 inches.

8.54. *Tulips 'Ballade' (pink with white edging) and 'Queen of the Night' (dark purple), along with daffodils (*Narcissus *'Thalia'), in flower in early May. (Photo: Robin Carlson)*

Tulipa 'Maureen'

Ivory-white flowers on tall stems, in May. 24 inches.

Tulipa 'Purissima'

Ivory flowers in Apri1.18 inches.

Tulipa 'Queen of the Night'

Deepest purple, almost black, flowers have made this a deservedly popular plant. Flowering May. 24 inches.

8.55. ***Tulip 'Ivory Floradale' emerges from a carpet of the bronze young foliage of*** **Astilbe chinensis** ***var.*** **taquetii** ***'Purpurlanze' in early May. (Photo: Robin Carlson)***

Tulipa 'Spring Green'

"The best tulip here, the flowers last longer than any of the others," says Colleen. White flowers forming a broad shallow bowl shape, each petal with a distinct green band on the outside. One of the best for year-to-year repeat flowering.
May flowering. 20 inches.

The following are known as "species" tulips, that is, they are descended from a single species, and in most cases are very similar to their wild ancestors. They are usually relatively short, with a wide range of flower shapes.

Tulipa bakeri 'Lilac Wonder'

Open, star-shaped, yellow-centered lilac-pink flowers, late April to early May. 16 inches.

8.56. ***Tulip 'Spring Green' has proved the most reliable in flowering from year to year in the Lurie Garden. (Photo: Robin Carlson)***

Tulipa *hageri* 'Splendens'

Open red flowers with a black base in late April, three to five to a stem. 16 inches.

Tulipa *polychroma*

White flowers with yellow centers having an attractive gray-purple flush on the outside. April flowering. 6 inches.

Tulipa *turkestanica*

Another species with yellow-centered white flowers—twelve to a plant, and distinctly star shaped. One of the most successful in the Lurie Garden. April flowering. 12 inches. Can spread through self-seeding.

8.57. **Tulipa bakeri** *'Lilac Wonder' is a so-called species tulip and is much closer in appearance to the wild tulips of central Asia than the familiar tall garden tulips. They are ideal for sunny situations on dry soils but do need high levels of nutrition if they are to flower again in future years—tomato feed is much appreciated! (Photo: International Flower Bulb Centre)*

8.58. **Tulipa turkestanica** *is very successful in the Lurie Garden. In some gardens, it can produce seedlings and spread. (Photo: International Flower Bulb Centre)*

Tulipa urumiensis

Yellow star-shaped flowers with green or red streaks on the reverse of the petals. 16 inches.

Tulipa wilsoniana

Vivid scarlet bowl-shaped flowers on 8-inch. stems. Glaucous leaves.

Resources

BOOKS

Piet Oudolf

Designing with Plants, Noel Kingsbury and Piet Oudolf, Timber Press, 1999.
Both an introduction to Piet Oudolf's planting design style and a beginner's guide to putting plants together in the garden.

Planting Design—Gardens in Time and Space, Noel Kingsbury and Piet Oudolf, Timber Press, 2005.
For those who already have some garden-making experience and for professionals, a more in-depth look at both Oudolf and other contemporary naturalistic design styles.

Landscapes in Landscapes, Noel Kingsbury and Piet Oudolf, Monacelli Press, 2011.
A photographic monograph of Oudolf design projects, public and private, around the world.

Planting—A New Perspective, Noel Kingsbury and Piet Oudolf, Timber Press, 2013.

Perennials

Armitage's Garden Perennials, Allan M. Armitage, Timber Press, 2011 (second edition).
The ultimate guide for perennial growers in North America.

Encyclopedia of Perennials, Graham Rice (editor), Dorling Kindersley, 2006.
Perhaps the most comprehensive, for the dedicated plant lover rather than beginner or designer, rather British focused.

Perennials: The Gardener's Reference, Susan Carter, Carrie Becker, and Bob Lilly, Timber Press, 2007.
Comprehensive without being overwhelming.

The Well-Tended Perennial Garden: Planting and Pruning Techniques, Tracy DiSabato-Aust, Timber Press, 2007 (second edition).
An innovative and thoroughly researched guide to managing perennials, particularly in smaller gardens.

The Know Maintenance Perennial Garden, Roy Diblik, Timber Press, 2014.
An introduction to simple planting formulas for the novice gardener.

Landscapes for Life: Combining Beauty, Biodiversity and Function in Garden Design, Rick Darke and Doug Tallamy, Timber Press, 2014.
Two leading advocates of naturalistic and native planting address the technical and aesthetic issues.

Perennials for Midwestern Gardens: Proven Plants for the Heartland, Anthony W. Kahtz, Timber Press, 2008.
One of a number of books aimed at regionally appropriate plants.

Natives/Wildflowers

The New England Wildflower Society Guide to Growing and Propagating Wildflowers of the United States and Canada, William Cullina, Houghton Mifflin, 2000.
Thorough and surprisingly witty, by a practitioner with huge experience.

Gardening with Prairie Plants: How to Create Beautiful Native Landscapes, Sally Wasowski, University of Minnesota Press, 2002.
Explains what prairie is, all the different variations, as well as being a practical guide to making prairie habitats at home.

Prairie Directory of North America, Charlotte Adelman and Bernard Schwartz, Lawndale Enterprise, 2001.
Nothing beats seeing the real thing—keep this one in the car!

ORGANIZATIONS

Chicago Wilderness. A regional alliance for protecting nature, with a magazine and informative website. www.chicagowilderness.org.

Citizens for Conservation. Actively involved in the conservation and restoration of wild areas in the wider Chicago region. Plenty of scope for volunteer involvement. www.citizensforconservation.org.

Chicago Botanic Garden. A public garden with a wide range of perennials, a large area of recreated prairie, and an extensive education program. www.chicagobotanic.org.

Morton Arboretum. The Schulenberg Prairie is one of the oldest prairie restorations in the Midwest. Educational program. www.mortonarb.org.

WEBSITES

The Lurie Garden, at http://luriegarden.org, has a comprehensive plant list, blog, and information about events.

Chicagoland Gardening, at www.chicagolandgardening.com, is a website for the region's leading garden magazine.

The Missouri Botanic Garden has a website that is extremely useful and comprehensive, functioning as an online plant encyclopedia. www.mobot.org/gardeninghelp/plantfinder.

Piet Oudolf has information about some of his projects, and plans, on his own website. www.oudolf.com.

Acknowledgments

As a Brit writing about a garden in the middle of the United States, I have sometimes felt a bit like a fish out of water, but then I tell myself that they got a Dutchman (Piet Oudolf) to design the planting. I know the plants (I have grown most of them), I know him, but to know and understand better the very different growing conditions of the American Midwest I have been dependent on the advice of Roy Diblik and Colleen Lockovitch. Roy runs the plant-growing side of a nursery and landscape business in southern Wisconsin—Northwind Perennial Farm—the garden he and his colleagues have made there is very special and every bit as much an inspiration as the Lurie Garden. Colleen, until she decamped to Portland, Oregon, muttering, "I can't stand another Chicago winter," was the Lurie Garden's chief horticulturalist, later director, and not only knows the plants and the climate intimately, but has a wonderful turn of phrase in describing them.

Jennifer Davit, who took over as director from Colleen in 2010, has been an immense help, in answering various questions and helping with access to photography. I would also like to thank Laura Young and Sylvia Schmeichel for garden information too. None of this would have been possible without the help of garden designer Piet Oudolf who has also kindly contributed many of the pictures. Thanks also to landscape designer Kathryn Gustafson for reading through relevant sections of the manuscript, and to Terry Guen for further information on the construction of the garden.

Photographs have come from a variety of sources. I would like to thank Robin Carlson, employed by the garden to document it through the year, many of whose images I have used extensively, and which convey the combination of natural beauty and public use that makes the garden special. Thanks also to Linda Oyama Bryan for providing some additional, very special, images. Also Sally Ferguson of the International Flower Bulb Centre, for help in providing pictures; for pictures of wild landscapes around Chicago, to Tom Vanderpoel and James Bodkin; and to Rick Darke for the picture of the garden's designer Piet Oudolf.

Finally, thanks to Christie Henry, Abby Collier, Mark Reschke, and the staff of the University of Chicago Press for seeing the value of doing this book and seeing it through.

Common Names and Scientific Names

Plant species and varieties are to be found alphabetically under their scientific name in the "Plant Directory" (chapter 8), as they are on the plant list on the Lurie Garden website: www.luriegarden.org/plantlife-list. The following list cross-references the common names most frequently used for the plants in the Lurie Garden.

PERENNIALS

alexanders: see *Zizia aurea*
anemone, Japanese: see *Anemone x hybrida*
aster, aromatic: see *Symphyotrichum oblongifolius* 'October Skies'
aster, New England: see *Symphyotrichum novae-angliae* 'Violetta'
aster, Tatarican: see *Aster tataricus* 'Jindai'
aster, white wood: see *Eurybia divaricata*
astilbe: see *Astilbe chinensis* var. *taquetii* 'Purpurlanze'
barrenwort: see *Epimedium* species
beard tongue, foxglove: see *Penstemon digitalis* 'Husker Red'
bee balm: see *Monarda bradburyana*
bergamot: see *Monarda bradburyana*
betony: see *Stachys officinalis*
bishop's hat: see *Epimedium* species
blazing star: see *Liatris spicata*
blue stars: see *Amsonia* species
bowman's root: see *Gillenia trifoliata*
burnets: see *Sanguisorba* species
butterfly weed: see *Asclepias tuberosa*
calamint: see *Calamintha nepeta* subsp. *nepeta*
catmint: see *Nepeta* species
compass plant: see *Silphium laciniatum*
coneflowers: see *Echinacea* species
coral bells: see *Heuchera villosa* 'Autumn Bride'
cranesbill: see *Geranium* species
Culver's root: see *Veronicastrum virginicum*
day lily: see *Hemerocallis* species
elcampagne: see *Inula magnifica* 'Sonnenstrahl'
eryngo: see *Eryngium bourgatii*

false Solomon's seal: see *Maianthemum racemosa*
false spikenard: see *Maianthemum racemosa*
foxglove: see *Digitalis ferruginea*
gentian, bottle: see *Gentiana andrewsii*
goatsbeard: see *Aruncus* 'Horatio'
goldenrod: see *Solidago rugosa* 'Fireworks'
hairy alum root: see *Heuchera villosa* 'Autumn Bride'
hedgenettle: see *Stachys officinalis*
hosta: see *Hosta* species
hyssop, giant: see *Agastache* 'Blue Fortune'
ironweed: see *Vernonia lettermanii* 'Iron Butterfly'
Jerusalem sage: see *Phlomis tuberosa* 'Amazone'
joe-pye weed: see *Eupatoriadelphus maculatus* 'Gateway'
knautia: see *Knautia macedonica*
knotweeds: see *Aconongon* 'Johanneswolke' and *Persicaria amplexicaulis* 'Firedance'
lead plant: see *Amorpha canescens*
lenten rose: see *Helleborus x hybridus*
loosestrife: see *Lythrum alatum*
maple leaf alum root: see *Heuchera villosa* 'Autumn Bride'
meadow rue: see *Thalictrum* 'Elin'
milkweed: see *Asclepias incarnata*
mountain mint: see *Pycnanthemum muticum*
old man's whiskers: see *Geum triflorum*
oregano: see *Origanum laevigatum* 'Herrenhausen'
peony: see *Paeonia lactiflora* 'Jan Van Leeuwen'
plumbago: see *Ceratostigma plumbaginoides*
poppy, oriental: see *Papaver orientale* 'Scarlet O'Hara'
prairie clover, purple: see *Dalea purpurea*
prairie smoke: see *Geum triflorum*
queen of the prairie: see *Filipendula rubra* 'Venusta Magnifica'
rattlesnake master: see *Eryngium yuccifolium*
rodgersia: see *Rodgersia pinnata* 'Superba'
Russian sage: see *Perovskia atriplicifolia* 'Little Spire'
sages: see *Salvia* species
sea lavender: see *Limonium platyphyllum*
shooting star, *Dodecatheon meadia* 'Aphrodite'
skullcap, hoary: see *Scutellaria incana*
soft shield fern: see *Polystichum setiferum* 'Herrenhausen'
Solomon's seal: see *Polygonatum biflorum*
speedwell: see *Veronica longifolia* 'Eveline'
stonecrops: see *Sedum* species

thistle, globe: see *Echinops bannaticus* 'Blue Glow'
toad lily: see *Tricyrtis*
twinleaf: see *Jeffersonia diphylla*
Virginia bluebells: see *Mertensia virginica*
wild ginger: see *Asarum canadense*
wild indigoes: see *Baptisia* species
wild petunia: see *Ruellia humilis*
wild quinine: see *Parthenium integrifolium*
wood avens: see *Geum rivale* 'Flames of Passion'

GRASSES

autumn moor grass: see *Sesleria autumnalis*
eulalia grass: see *Miscanthus sinensis* 'Malepartus'
fountain grass: see *Pennisetum alopecuroides* 'Cassian'
Indian grass: see *Sorghastrum nutans* 'Sioux Blue'
little bluestem: see *Schizachyrium scoparium* 'The Blues'
love grass, purple: see *Eragrostis spectabilis*
moor grass: see *Molinia caerulea*
panic grass: see *Panicum virgatum* 'Shenandoah'
prairie dropseed: see *Sporobolus heterolepis*
reed grass, feather: see *Calamagrostis x acutiflora* 'Karl Foerster'
reed grass, Korean feather: see *Calamagrostis brachytricha*
sea oats, northern: see *Chasmanthium latifolium*
sedges: see *Carex* species
switchgrass: see *Panicum virgatum* 'Shenandoah'

BULBS

camas: see *Camassia* species
crocus: see *Crocus* species
daffodil: see *Narcissus* species
fritillary: see *Fritillaria* species
glory of the snow: see *Chionodoxa forbesii* 'Blue Giant'
grape hyacinth: see *Muscari* species
Grecian windflower: see *Anemone blanda* 'Blue Shades'
Indian blue hyacinth: see *Camassia* species
ornamental onions, garlics, leeks: see *Allium* species
quamash: see *Camassia* species
squill: see *Scilla mischtschenkoana*
star of Persia: see *Allium christophii*
tulips: see *Tulipa* species

Index

Ann and Robert H. Lurie Foundation, 5
annuals, 19
Art Institute of Chicago, 3, 7
aspect, 18, 45, 48

bees, 4, 84–85, 144
biennials, 19, 20, 106
bulbs, 15, 17, 19, 21–23, 41, 49, 54, 59, 65–68, 73, 143, 168, 171, 176–95
butterflies, 81, 83–84, 85, 90, 98, 113, 144, 155, 158, 162

climate, 16–28, 35, 82, 87. *See also* microclimate
clonal and nonclonal perennials, 20–21
color, in gardening plantings, 10, 12, 14, 22, 25–26, 54, 59, 62–63, 66

Dark Plate, 3, 6, 12, 22, 45, 112, 118, 148
deadheading, 70
deer, 43–44, 64, 98
design: issues in gardens and planting, 53–64; of the Lurie Garden, 6–12
Diblik, Roy, 12–13, 197
DiSabato-Aust, Tracy, 70, 197
diseases, fungal and bacterial, 2, 24, 25, 37, 39, 40, 42
division (propagation), 20, 26, 67

fall, and fall colors, 25–26, 54, 58, 60, 62, 66, 73, 76, 91, 92, 93, 95, 102, 103, 110, 116–17, 122–23, 125, 136, 140, 141, 147, 148, 153, 154, 156, 171–76
fertilizer. *See* nutrients, for plants

grasses: ornamental species, 7, 10, 12, 13, 24, 26, 44, 50, 58, 59, 61, 65, 66, 73, 75, 164–76; in prairie habitats, 23, 75–79
Guen, Terry, 12, 13
Gustafson, Kathryn, 5–7

habitats, wild, 4, 7, 12, 30, 38, 75–80
hardiness, of plants in winter, 17, 22, 26, 87
hedges, the Shoulder Hedge, 3, 7

irrigation, 24–25, 30, 32, 37, 39–42, 70–71
Israel, Robert, 5

latitude, impact on garden plants, 46–47, 87
lifespan, of garden plants, 19–22, 50–51, 191
Light Plate, 3, 6, 7, 10

maintenance, 31, 49–50, 53, 64, 67–74
Meadow, The, 10, 12
microclimate, 18, 48
moisture content of soils, 25, 30–34, 37, 40, 47–48
mulches and mulching, 33, 37, 38–39, 70, 73

native plants, 2, 3–4, 12, 22, 24, 31, 33–34, 45, 75–85
nutrients, for plants, 29, 31–37, 38, 39, 46, 47, 80

organic matter, as soil improver, 29, 31, 32, 37, 38, 39, 46
Oudolf, Piet, 2, 5–14, 26, 54, 59, 61, 63, 64

perennials (discussion of meaning), 1, 19–20
pests, 2, 3, 42–44, 71–72
planting: design, 55–66; styles, 10, 13–15, 29, 30, 42; as a task, 23, 28, 37, 43, 67–68
planting companions, 61–62
prairie habitat, 23, 45, 63, 75–81
propagation, 20, 51, 67
pruning, of perennials, 70

rabbits, 43, 90, 105, 110, 134, 146, 159, 162

Salvia, European meadow species, and Salvia River, 3, 10, 18, 70, 152
seed (propagation), and self-seeding of garden plants, 19, 20, 50–51, 52, 73–74
seed heads, 25, 26, 54, 60, 62, 66, 73, 83
shade, 6, 12, 45–47, 49, 61, 64, 87. *See also* woodland perennials
spread, as aspect of perennial growth, 2, 20–21, 50–53, 59, 73, 86–87
spring, 17, 21, 22–23, 27, 48, 49, 54, 66, 67–70
soil and soil types, 10, 25, 28, 30–33, 35, 37, 38, 47, 61, 67–68, 79, 87
staking, of perennials, 27, 36, 70
summer, 10, 16, 17, 22, 23–25, 27, 31, 36, 47, 54, 58, 59, 63, 66, 70–73, 75, 84

tulips, 22–23, 49, 67, 68, 164

watering. *See* irrigation
waterlogging, of soils, 32, 34, 48, 62
weeds, 17, 29, 37, 38, 43, 51, 69–70, 72–73, 83
wetland, 21, 34, 35, 48, 79
wildlife, 3, 26, 73, 79, 81–85
wind, 16, 17, 26, 27–28, 36, 48, 70, 72
winter, 16–17, 23, 25–27, 38, 50, 54, 59, 60, 65, 66, 67, 73–74, 83
woodland perennials, 19, 22, 33–34

zones, USDA (for plant hardiness), 17–18, 82, 87